ENGINEERING LIBRARY

Mobile 3D Graphics

Alessio Malizia

Mobile 3D Graphics

With 54 figures

Alessio Malizia
University of Roma
"La Sapienza"
Rome
Italy

British Library Cataloguing in Publication Data
A catalogue record for this book is available from the British Library

Library of Congress Control Number: 2006924370

ISBN-10: 1-84628-383-3
ISBN-13: 978-1-84628-383-3

Printed on acid-free paper

© Springer-Verlag London Limited 2006

Apart from any fair dealing for the purposes of research or private study, or criticism or review, as permitted under the Copyright, Designs and Patents Act 1988, this publication may only be reproduced, stored or transmitted, in any form or by any means, with the prior permission in writing of the publishers, or in the case of reprographic reproduction in accordance with the terms of licences issued by the Copyright Licensing Agency. Enquiries concerning reproduction outside those terms should be sent to the publishers.

The use of registered names, trademarks, etc. in this publication does not imply, even in the absence of a specific statement, that such names are exempt from the relevant laws and regulations and therefore free for general use.

The publisher makes no representation, express or implied, with regard to the accuracy of the information contained in this book and cannot accept any legal responsibility or liability for any errors or omissions that may be made.

Printed in the United States of America (MVY)

9 8 7 6 5 4 3 2 1

Springer Science+Business Media
springer.com

In memory of
Enzo Luigi Malizia,
Father and Guide

Preface

We may consider technology as the set of tools, both hardware and software, that help us improve our performance in different working (and playing) locations; it includes all the man-made objects from paper to the latest audio book. Information technology helps to use, edit, share, and exchange knowledge, in the form of documents— textual, acoustic, and pictorial—quickly and efficiently. E.D. Hirsch, Jr. pointed out that literate people in every society and every culture share a body of knowledge that enables them to communicate with each other and make sense of the world around them. The kinds of things a literate person knows vary from society to society and from era to era, so there is no absolute definition of literacy; the same holds for computer literacy. We may look at technology literacy from three different dimensions: capabilities, knowledge, and ways of thinking and acting. According to this scheme, such dimensions can be placed along a continuum - from low to high, limited to extensive, and poorly developed to well developed. In such a three-dimensional (3D) space, are the different products and gadgets in our everyday life, like the iPod, that are extensively developed, and have high-level capabilities but require little knowledge to make them work. Conversely, a CAD software package has low capabilities since it is a specialized application, and extensive knowledge is required to use it. These ways of thinking and acting must be well developed. There are different views on which computing platform will encompass others existing one. At first the office computer substituted for the mainframe computer. Next came the laptop, which worked even better than the desktop, but its short battery life still remains a shortcoming. Personal digital assistants (PDAs) and pocket computers are rivaling laptops in providing services that were only possible on larger machines. Next came the cellular phone with its color screen, its variety of capabilities and its connection to the Internet. Even on this small and light device, the video games (first played only on TV screens) are gradually expanding with respect to the home entertainment area, but, at the same time, they are also becoming communication platforms employing their processing power not only to create characters and their corresponding animation but also to work with email and to exchange messages. The question remains: How will people change their behaviors in order to use new services on the cellular phone, modifying their day-to-day habits so that the new portable phone may next become a real media center? We all are resistant to changing our habits; it is an inborn inertia that stops us from learning new systems. Such an initial learning phase requires extra time, but it leads us to the efficient use, in this case, of a multi-functional cellular phone. This quickly evolving scenario makes it difficult to forecast when and how we will be using new kinds of cellular phones in the near future. Nevertheless, their graphics possibilities, memory capability, and higher computing power will surely enable them to display complex, animated pictures, which may represent different kinds of useful information, as will be extensively demonstrated by this book. For instance, by looking at the phone screen, we could take a quick glance at the stock market and graphically see the trends at other international markets. We could see weather forecasts, which could be displayed by animated icons portraying rain or sunshine with a false-color temperature scale, or even watch short videos or film trailers. We may therefore assume that the cellular phone will become something very close to a pocket computer that also has audio communication facilities so that we can keep calling it a phone! Nevertheless some usability problems will crop up: How small can the number buttons be? How large must the display be in order to enable any user to see it clearly? How much battery independence should we have, so as to let the user forget about power consumption and recharging? Indeed, small is beautiful, as many people have said in the past, but too small can be disappointing. This book begins with the description of a set of applications in mobile graphics (graphics as performed on a cellular phone), including possible scenarios and relative challenges. The details of 3D graphics using the OpenGL®ES libraries are next provided together with their possible extensions in the future, followed by the Java™Mobile 3D graphics and Direct3D®Mobile libraries to perform similar functions. Finally, some samples and cases are shown to better illustrate the capabilities of the above software libraries. After reading this book, you will be up-to-date in the realm of 3D graphics executed on the cellular phone of the future. We are betting such a miniplatform will enable the user to reach, a large amount of services and information that were previously available only through other, larger, platforms.

Stefano Levialdi
Life Fellow, Institute of Electrical and Electronics Engineers
Rome, Italy
March 2006

Acknowledgments

Many people have contributed to this book in a variety of ways over the years. To the organizations and individuals who provided pictures and other materials, I express all my appreciation. I also want to acknowledge the many helpful comments received from our students in various computer-graphics and interaction courses, especially: Neri Vicari, Elisabetta Favilene, Angelo Moriconi, Adriano Rinaldi and Francesco Colonnese. I'm very obliged to all those provided comments, reviews, suggestions for improving the materials included in this book, and I want to extend my apologies to anyone may have forgot to mention. My thanks to Prof. Stefano Levialdi, for his precious help and the preface; Prof. Steven Tanimoto for his support at every level, and all the HCI group at University "La Sapienza" of Rome. I want to mention my family for the crucial support given to me during this work: my lovely wife, Claudia, my mother Maria, and my brother, Davide. Thanks also to all my dear friends and especially to Andrea Calabresi and Paolo Trevisanutto. Many thanks to the Springer editors and staff who supported me in writing this book: Beverly Ford, Helen Desmond, Joanne Cooling, Varghese Kozhimannil, Lesley Poliner; I want to extend my sincere appreciation for their many competent contributions and careful effort to detail.

Contents

Preface vii

Part I Scenarios and Applications

Mobile Graphics Applications 3
1.1 The Rise of Mobile Graphics 3
1.2 Mobile Devices 4
1.2.1 Platforms 5
1.2.2 Software 7
1.2.3 Usability 8
1.3 Mobile Devices and Graphics 11
1.3.1 Graphics Software 11
1.3.2 Rendering Pipeline 13
1.3.3 Wire Frame 16
1.3.4 Depth Cuing 16
1.3.5 Hidden Surfaces 17
1.3.6 Lighting Models 20
1.3.7 Textures 21
1.3.8 Mobile Graphics Software 22
1.4 Summary 27

Mobile 3D Graphics: Scenarios and Challenges 29
2.1 Application Scenarios 29
2.2 Multimedia and Graphics Usability Challenges 39
2.3 Algorithms and Architectural Challenges 41
2.3.1 Fixed-Point Maths 44
2.3.2 Graphics Hardware Architecture Design 46
2.3.3 Tile Rendering 49
2.4 Summary 52

Part II Mobile Graphics Programming

Introduction to Mobile 3D Graphics with OpenGL®ES 55
3.1 Introduction to OpenGL®ES 55
3.2 The OpenGL®ES Rendering Pipeline 56
3.3 3D Mobile Graphic Concepts and Rendering with OpenGL®ES 56
3.3.1 Starting with a Window 56
3.3.2 Basic Interaction 59
3.3.3 Geometric Primitives and Per-Vertex Operations 59
3.3.4 Lighting 68
3.3.5 Per-Pixel Operations and Texture Mapping 75
3.3.6 Per-Fragment Operations 78
3.4 OpenGL®ES Future Developments and Extensions. 82
3.5 Summary 84

Java™Mobile 3D Graphics 86
4.1 M3G 86
4.2 MIDP Applications 88
4.3 Immediate and Retained Mode 91
4.4 Scene Graph 92
4.5 Transformations 94
4.6 Nodes of the Scene Graph 94
4.7 Camera Class 95
4.8 Managing Illumination 95
4.9 Meshes and Sprites 96
4.10 Animations 98
4.11 Ray Intersections 100
4.12 Building an M3G Demo 100
4.12.1 DemoCarCanvas Class 103
4.12.2 Car Class 109
4.13 Summary 115

Direct3D®Mobile 116
5.1 Architecture 116
5.2 Rendering Pipeline 118
5.3 Primitive Types 119
5.4 Transformations 122
5.5 Lighting 126
5.6 Summary 128

Conclusions and Prospect 129

Appendix A: OpenGL® ES Code Samples 131
A.1 Starting with a Window 131
A.2 Basic Interaction 134
A.3 Geometric primitives and Per-Vertex Operations 135
A.4 Lighting 143

References 153

Index 157

Part I

Scenarios and Applications

1

Mobile Graphics Applications

1.1 The Rise of Mobile Graphics

In the last few years we have seen a dramatic growth in both the computation and connection capabilities of mobile computing devices. Today, all this power can be packed into a device so compact that it fits in your pocket. In fact you can go out your front door carrying your favorites images, videos, and data files, working wherever you like [1]. Throughout this book we will talk about mobile devices including palm-tops and mobile phones, which are very compact with limited amount of RAM (random access memory). We will also talk about handhelds, with still compact but wider screens than mobile devices and more RAM and CPU power. Not only has the computation and communication power of mobile devices and handhelds been increased, but the visualization and graphics capabilities have also risen. You can now take a picture or capture a video sequence with your cellular phone or personal digital assistant. Moreover, you can view as many colors as on a graphic workstation and interact with color graphics applications in real time. Even considering these major advancements in mobile devices, there are still unresolved problems. Manufacturers can integrate all the multimedia content they want, but if the battery runs out and users can't, for example, make a brief phone call, the tool will not be effective. The challenge is to integrate all this video and multimedia technology without sacrificing the battery life of the product or weighing it down with expensive batteries. There have been advancements in microprocessor management of battery power consumption and the graphics capabilities associated with it, but with these enhancements the desktop computer continues to be the leader in providing multimedia experiences to end users. However, this may change very soon.

1.2 Mobile Devices

The first question to ask is: Which mobile applications will best suit the new generation of graphics capabilities and user needs? Entertainment, games, computer-aided design (CAD), virtual reality (VR), data visualization, and education and training are all examples of application areas that drove the development of graphics on desktop computers. Some of them, such as CAD or VR, even if they could be supported by the new handheld computers and mobile devices, might not be suited in terms of ergonomics [2]. It's hard to imagine a designer developing a new car on a mobile device even with new mobile hardware capabilities. Even considering the new graphics features, mobile devices are still less capable than desktop computers. In fact, programs run at lower speed, there is less memory for processing and storing programs, the displays are smaller in size and color numbers, and the battery could run out. The most important reason behind these differences is power: one device is plugged in to the power grid, while the other is dependent on battery power. Power management is of extreme importance to manufacturers, and for everyone in this market including content providers, handset vendors, and carriers. However, mobile devices will improve. The continuing exponential reduction in size and cost (Moore's law) will allow more processing power, memory, and storage for mobile devices. Also, batteries will last longer with advancements in integrated circuit power consumption. Moreover, mobile devices are already superior to the desktop computers of 10 years ago. In particular, the computer graphics and visualization capabilities on these devices are improving quickly.

Another relevant issue to deal with is the usability of mobile devices. In fact, manufacturers have reduced the scale of devices, and are addressing the difficulties in using handhelds. Many of them have problems with inputting data (keyboards are too small or pen-based systems require the user to learn a specific language), or provide miniaturized displays and very small visual interfaces. The following problems can be distinguished:

- storage and data size limitations,
- communications and graphics interfaces usability problems.

It is important to think about data size reduction in order to produce an acceptable real-time visualization. There are basically two approaches to data reduction:

1. the *generic* approach consists of data structures and size reallocations to match the device limitations, without considering user needs and processes;
2. the so-called *specific* approach provides the minimum requirements for executing the user processes.

For instance, in the generic approach the original text of an article could be synthesized by small paragraphs representing its sense, while images and videos could be coded with less resolution in order to reduce size and space.

Communications problems deal mainly with wireless networks communication. Wireless communication requires significant power consumption and could easily discharge batteries of mobile devices. In case of power or energy loss, the applications should automatically save their state (a snapshot of critical system resources) from time to time to allow recovery after a power shut-down. Graphics interfaces provide interesting functionalities and features on mobile devices; in fact, the user can interact directly with the screen via pen-based software or touch sensitive screens. Visual and graphics interfaces for mobile devices basically use three approaches. The first consists of one handheld the device, and the user's second hand interacts with the visual interface via a pen device. The second approach is to have special buttons on the device that help users in performing certain task. The third approach consists of using a touch-sensitive screen and fingers or stylus devices to interact with the applications. This last method has started to be supported by many manufacturers because of the decreasing costs of touch screens and the fact that fingers could be intuitively used.

1.2.1 Platforms

Graphics workstations have existed for the last 20 years, but many of the design choices made for them are not suitable for mobile devices. In fact, they require too much power, expense, and space to be cloned on mobile devices. However, some approaches, like video cards supporting OpenGL application programming interfaces (API), can be adapted with fewer features, maintaining the core functionalities. User expectations, which were set by desktop computers, will push manufacturers to fill the gap in graphics capabilities between mobile and desktop computers. There is no need for completely new ideas, though creative approaches are always welcome. The main problem is how to fit graphics workstation features into mobile devices.

Handheld devices (Figure 1.1) vary widely in capability, ranging from 400-MHz PDAs with 64 MB RAM, to 50-MHz mobile phones with 1 MB RAM. Graphics applications have to be designed to accommodate these differences by enabling small implementations with minimum data storage requirements, minimized instruction/data traffic, and so on. For users, this means smaller binaries to download that consume less storage on the device. Moreover, there are different kinds of hardware. Let's consider the following classifications, which will help us in identifying mobile device platforms:

- **Handheld Computers**
  These are small, light, and fit into pockets. They can be connected to the Internet, and users can usually input data or run applications via a pen and a touch screen. Usually they provide icons and buttons on the device that help users to quickly run frequently used applications.
- **Mobile Phones**
  Modern mobile phones, thanks to new technology enhancements, have evolved from using a voice-based interface (phone calls being the main

application) to having powerful network clients. There are many different mobile phones on the market today: Java™ based, including photo and video cameras, supporting UMTS (universal mobile telecommunications system) and Bluetooth, thousands of color graphics displays, and so on.

- **Smart Phones**
  Smart phones are a combination of mobile phones and handhelds with an organizer in a single communication system. Smart phones usually allow wireless connections supporting faxes, e-mail, SMS (short message service), Internet access, applications, and Personal Information Management (PIM) software. They can also be easily connected to a PC via USB cables, wireless interfaces, Bluetooth, or infrared connections.

For graphics applications, the industry is setting up optimization processes in order to fit all the features into mobile and handhelds devices. Graphic and multimedia content will stimulate the development of mobile applications only if they respect the desktop standard people are accustomed to. Hardware and software development follows user needs but also requires investment in terms of money and time. To fit the graphics requirements and meet cost requirements, standards should be developed and adopted in developing mobile graphics applications.

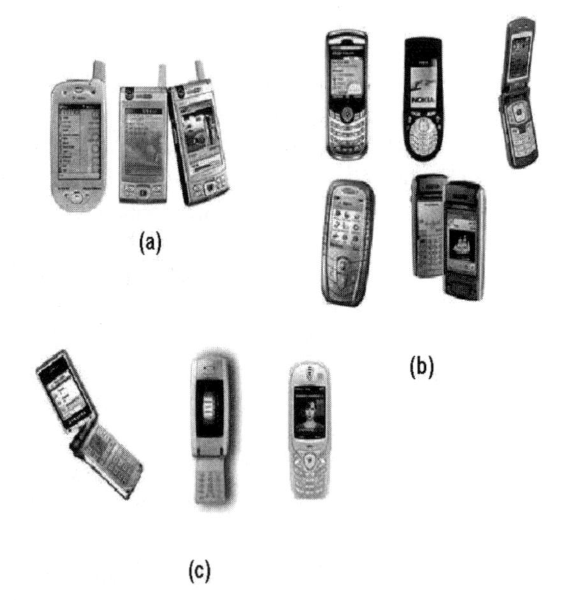

Fig. 1.1. (a) Handhelds devices. (b) Mobile phones. (c) Smart phones.

1.2.2 Software

Since we will see how to design and develop three-dimensional (3D) applications for mobile devices, we need to know which software platforms are supported by handhelds in order to manage different operating system capabilities.

Mobile devicessupport different operating systems, and we will describe the three most widespread: Microsoft Windows®, Palm OS ®, and Symbian OS™.

Microsoft Windows for pocket devices started in 1996. The application level interface is based on API Thus developers would find a subset of classical Windows API (similar to the desktop version) and developer tools like Visual Basic®, Visual C++®, etc.

The user interface for pocket PCs is similar to other Windows-based personal computers. The Start menu is on the bottom of the display, as is the task bar; however the status bar has moved to the top. Every application is embedded into a window and can be managed from that window.

Palm OS has been developed by Palm™, Inc., and is a widely used handheld operating system. It includes programming API, and a good hardware abstraction layer that helps in porting to new platforms (different kinds of CPUs). It also has features concerning security, color displays, e-mail, and wireless Internet access.

Palm OS manages memory with so-called *memory cards*. A memory card is a logic unit made of RAM, read-only memory (ROM), or both. The entire main memory is split into different heaps: one single dynamic heap, and multiple storage heaps.

Palm OS is an event-driven OS; everything is based on events raised by the user or applications, caught by event handlers.

There are three major programming languages used for developing applications on Palm OS: C, C++, and Java. Because of the limited heap size, C++ is preferred to C++ and Java. Palm provides two packages for application development:

- The Palm Software Development Kit (SDK) includes API for programming interfaces, managing the system, and communications.
- The Conduit Development Kit (CDK) supports the conduits development for synchronizing and backing up data/applications to and from the Palm device and the desktop PC.

The last of the most used operating systems (OSs) for handhelds is **Symbian OS**, also known as EPOC®(EPOC was a range of operating systems developed by Psion™for portable devices, primarily PDAs). It is a real-time, 32-bit multitasking OS that uses C++ and an object-oriented approach. Symbian and EPOC were started by the PSion software company. In June 1998, PSion, together with Ericsson™, Motorola™, and Nokia™founded Symbian as a joint venture for developing wireless and mobile platforms. Later on, Panasonic™, Sony Ericsson™, and Siemens™also joined the Symbian venture. It is a stable OS for handheld and mobile devices. It runs on x86 PC processors and various ARM Ltd (advanced RISC machines) CPUs.

Symbian OS provides a three-layer model for developing applications. The three layers are the following:

- *Application Engine.* It includes the logical structure of the application. It is a model of the application steps.
- *Application View.* It provides a simple graphic view of the application data. Thus developers can abstract from specific graphics user interfaces GUI (different handhelds or mobile phones can render the same GUI item in different ways).
- *Application GUI.* It defines the possible graphic views to be displayed to the end user.

The Symbian OS has been designed and developed using C++, and all the API are developed with this language. The only way to access all the OS capabilities is by wrapping C++ code around the provided API. Symbian now supports also PersonalJava (now extended to J2ME™standard, and it is described later in this chapter), which is the reference platform for the JavaPhone® API.

After this brief introduction to mobile device platforms, features, and characteristics, in the next section we describe the graphics software and packages running on mobile devices, recalling the usability issues that will help us guide the development of mobile graphics applications from a user-centric viewpoint.

1.2.3 Usability

Marketing competition for size reduction has enlarged the difficulties in using mobile devices interfaces both for input (small "qwerty" keyboard) and for output (reduced screen size). Many handhelds and mobile devices on the market have small keyboards or pen-based symbolic languages that are not error-prone and need to be learned by users. Moreover, they usually include small screens, such that they affect readability and precision in pen-based interaction systems.

This section focuses on usability issues for multimedia and graphics applications dealing with dynamic data. Since the scope of this book is about mobile 3D graphics, we concentrate on this particular domain, which poses problems that influence all the various phases of graphics rendering. A study [3] that focuses on graphics and multimedia applications with mobile devices treats the usability issues by cataloging them as follows:

- Data reduction
- Data communication
- Graphic interfaces

It has to be noted that a data reduction process is needed in order to provide effective real-time visualization for a mobile device.

There are mainly two approaches: The *generic approach* consists of coding data for adapting to the mobile device limited resources, independently from the users' needs or task requirements.

The *specific approach*, instead, provides the minimum amount of data and information needed for completing the user tasks (which implies knowledge of users' task and priorities).

In the generic approach, for instance, the text could be summarized in paragraphs and sentences that contain the keywords and the real sense of the submitted text. Images and videos could be provided at lower resolutions (data reduction) and also audio resolution could be reduced.

In the specific approach, video resolution could be adapted depending on the user tasks, so, for instance, if the task is surveillance, a time interval could be set and only relevant frames within this time interval are rendered with high resolution; if the task is watching a sports event, the system provides lower resolution images but achieves high frame rates for displaying real-time actions.

Data communication problems are mainly related to the wireless capabilities of handheld devices. Wireless network support involves high power consumption, and thus batteries can run out of energy. In this case (low energy levels) the applications must be aware of the low power and perform all the necessary operations for data recovery, for example, saving data and system status. Some recovery procedures and techniques are listed in [4]. For example, one-to-one communications can be directly supported among individual mobile devices.

Graphics user interfaces (GUI) can support direct manipulation for enhancing virtual or augmented reality experience on mobile devices. By using these visual interfaces, users can manipulate and manage virtual objects in virtual space (useful not only for games but also for augmented reality application, as we will see in Chapter 2). Support for these virtual functions can be embedded in mobile devices using and supporting tactile and/or optical sensors. Some new research papers use the concept of physical and virtual spaces for multimodal communication on mobile devices. They allow users to manipulate, rotate, and move physical objects that reflect their position in user interfaces, thus allowing direct manipulation [5, 6]. Graphics interfaces for mobile devices employ two main approaches. The first approach consists of designing physical interactive objects separately from the design of mobile device and then using wireless connections to enable communications. In this way the dominant hand could handle direct manipulation (while the other hand holds the mobile device). This approach could involve some awkward motions because the same hand is in charge of object selection.

One solution for this problem could be found in designing buttons for these physical manipulation devices in smart locations; thus object selection is performed directly by using fingers. An alternative approach consists of reshaping devices and redesigning basic functionalities in order to embed more intuitive and advanced kinds of inputs.

The most often used input communication modality for visual interfaces is the *touch screen*, by recognizing the level of hand or finger pressure, or the stylus pointing at the screen. This technique, at its origin, was used as an alternative communication technique (playing sounds corresponding to touching actions on the screen) for visually impaired people. The main advantages of implementing this technique are the low cost of hardware support, and the minimal space requirements. Moreover, it not only can support finger interaction, but also other body parts could be used when needed (for impaired people or as an additional aid for generic users).

1.3 Mobile Devices and Graphics

1.3.1 Graphics Software

Graphics software can be split into two main categories: special purpose and general purpose. Special-purpose graphics software is oriented to end users, which means it provides a set of functions to draw pictures, graphs, animation, and so on. The user is usually an expert in his or her domain, but all the graphics libraries and functions needed to perform the graphics tasks are hidden from the user. Usually interfaces in these packages are made of menus and icons representing and summarizing the different graphics functions, such as lines, circles, color palettes, and so on. These special-purpose applications are used in many different fields including engineering CAD and computer-aided manufactoring (CAM) systems, architecture, business, medicine, and many others. There are many such application packages including Photoshop, AutoCAD, Maya, and 3DStudio. In other chapters we will see that in specific cases these special-purpose applications help us in working with our graphics applications. In fact, sometimes using applications like Maya to set up a 3D scene and then importing 3D data into a graphics application can result in a more realistic scene and faster development of the graphics application [7, 8].

General-purpose packages usually come in the form of API and libraries. They provide a set of graphics functions and integrate them with programming languages like C++ C++, and Java. The set of functions included in these packages usually includes geometric functions for lines, polygons, circles, ellipses, spheres, cubes, and others. They also provide different settings for color models and spaces, points of view and camera positioning, shading, transformation, and modeling functions. We are especially interested in the packages including OpenGL®, OpenGL®ES, VRML (Virtual Reality Modeling Language), Java™ 2D and 3D, and Microsoft DirectX®. Using these packages (we will focus on OpenGL/OpenGL-ES and Java 3D/J2ME/JSR184) users can write their applications in C++ C++, or Java and integrate these libraries and API, design a 3D scene, display it on a screen, and interact with it [9].

The basic API that a general-purposegraphics library can provide are usually for image creation and management. These functions are called *graphics primitives*. They usually provide functions for representing and drawing points, lines, curved lines, polygons, circles, and different kind of shapes defined by an array of points. They also include color-filling tools (for polygons and closed areas) and sometimes basic 3D shapes like cubes, spheres, and others. The graphics primitives also use attributes to specify colors, line styles, and filling patterns. In addition, we usually find API for *geometric transformations* such as changing the size, orientation, and position of a shape or a set of shapes. There are API for *viewing transformations*, which are useful for setting up a 3D scene. They provide functions for selecting the point of view (the camera position) and type of projections to be used (e.g., parallel or perspective). Other API are provided for *clipping*, in order to determine the visible surfaces, and *lighting* for specifying the source and kind of illuminations. There are API used for *input functions*, which manage the keys pressed by the users, interfaces, and the mouse. There are finally functions for initializing the screen and managing refresh rates and color palettes, usually called *control API*.

An important issue with geometric modeling is how to represent the geometric description of the objects to be displayed. A point usually needs two coordinates to be displayed, a rectangle is specified by its corner coordinates (two vertices, left-top and right-bottom), and a sphere by its center and a radius. Usually graphics packages use the Cartesian-coordinate system and if coordinates are expressed in a different reference space such as a sphere, then a coordinates conversion function is needed. The process of modeling a scene requires using different reference coordinates. In the first step we define the shape of our objects, drawing or modeling each object with its own reference space. These reference spaces are called **modeling coordinates**. We can build a model of the scene by placing each object in the scene reference space. This process involves transforming the modeling coordinates in the scene reference frame, the so-called **world coordinates**, where we can place, in different positions and orientation, the shapes of our modeled objects. For example we can build a robot, as shown in Figure 1.2, by defining each of its parts (base, upper body, arms, and thumbs) in modeling coordinates, and then join all the parts in the world coordinate space. We can define the arm once in the modeling space and then place two instances of the same object with different locations (close to the left and right side of the body) in world coordinates. For a better representation and a clearer model of how to assemble the modeling parts in the world coordinate space, we could use a structure like a directed acyclic graph (DAG), or a tree showing the connections between parts of the model. We could express our coordinates in whatever measure we need, so, for instance, for one scene, our coordinates could refer to meters, while in another it could be kilometers or miles. It is up to us to decide how to interpret the world coordinate locations.

After computing and transforming the scene in world coordinates, there is a process composed of various routines, where usually the output of one routine serves as the input to other routines, in order to transform the scene for different output devices. This process is called the **graphic pipeline**. First, a view of the scene is built converting the world coordinates in the **viewing coordinates**, corresponding to the position and orientation of a virtual camera. This process gives us the scene with respect to the desired point of view. Then the viewing coordinates are projected onto a 2D space, which is the one displayed by the graphics output devices (e.g., graphic cards and screens). After this step, the coordinates are normalized to lie in the range $[-1,1]$ or $[0,1]$ and depending on the kind of graphic device used. Moreover, we have to identify visible surfaces in order to properly display the scene, and remove the surfaces outside the selected view. Finally, the graphic system

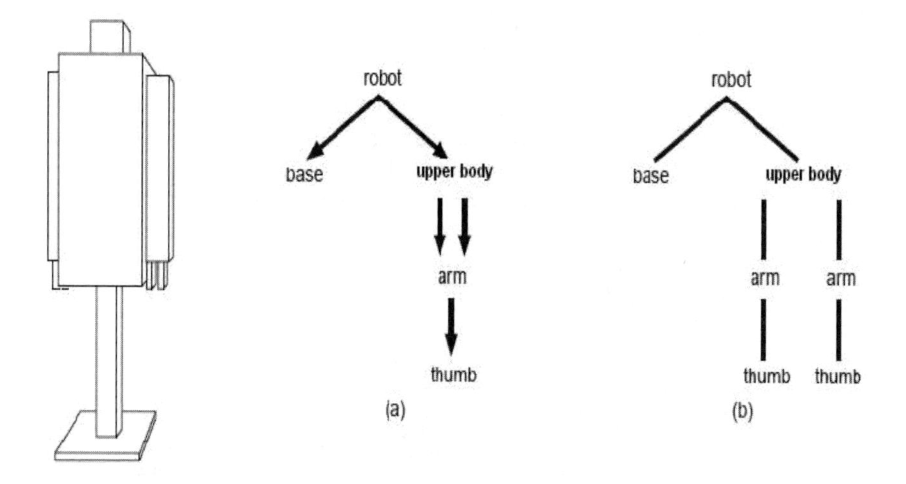

Fig. 1.2. Modeling an object by its parts. (a) The direct acyclic graph. (b) The tree associated with the model.

needs to have all the graphics information converted in the refresh buffer to actually be displayed. The coordinate system for the display devices are called **device/screen coordinates**. An example of a typical graphic pipeline is shown in Figure 1.3.

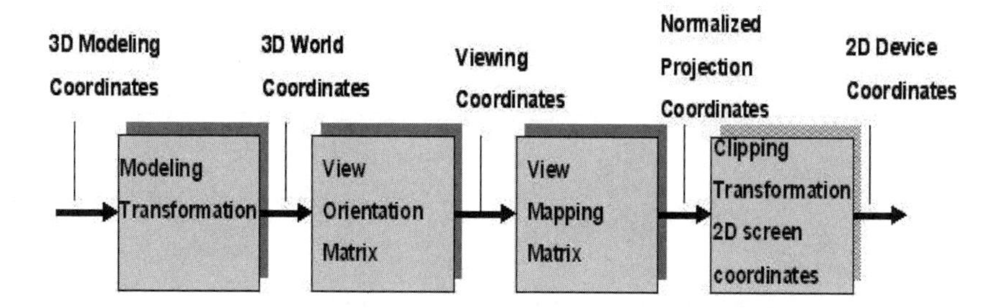

Fig. 1.3. The transformation pipeline from modeling coordinates to device coordinates projection for displaying a 3D scene.

1.3.2 Rendering Pipeline.

We will now focus on the steps needed to transform an object geometric description into an image rendered on the screen. This process is called *Rendering Pipeline*. The OpenGL pipeline structure is a clear and effective example of a rendering pipeline, and we will use it for explaining the basic concepts of rendering. OpenGL is structured as a finite state machine, and thus parameter values can be considered as states and language commands act as a transition changing the current state. For example, setting a color value to red (*set color*) will affect all subsequent instructions and geometric primitives until a new color is selected. This behavior holds for every primitive parameter (color, shading, transparency, lightning, material, etc.).

The OpenGL rendering pipeline includes: per-vertex operations and assembly primitives, per-pixel operations, rastering, per-fragment operations, and frame buffer operations.

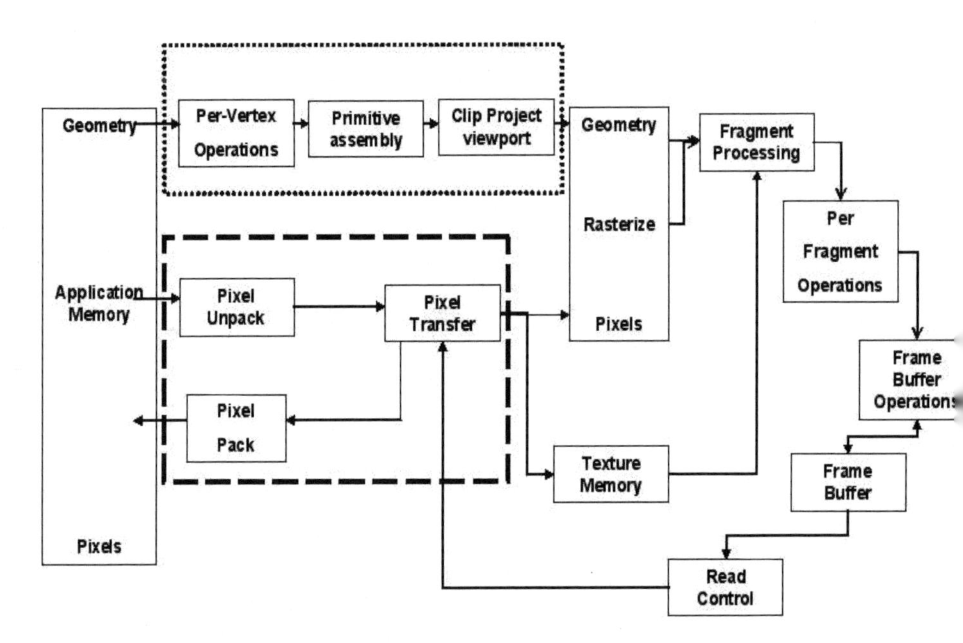

Fig. 1.4. OpenGL rendering pipeline.

The rendering process includes two types of input:

1. Spatial coordinates, which means vertexes and points
2. Images, pixels, and bit maps.

Geometric operations, so called *per-vertex* operations, convert the vertexes (Figure 1.4 small dashed rectangle) into geometric primitives. The lighting computations take place also in this phase, since the normal vectors with respect to the vertexes are needed in order to compute light intensity and direction. Moreover, the light effect onto the materials is computed to render a photo-realistic scene.

Per-pixel operations (Figure 1.4 big dashed rectangle) compute the pixel transformations, as initially images are stored in memory as an array of pixels.

The two processes, per-vertex and per-pixel, are then combined together in the rasterization process, and both vertexes and images (pixels) are transformed into *fragments*. Each fragment represents a pixel in the frame buffer. The frame buffer holds the image frame that will be displayed on the screen.

The *per-fragment*operations will fill polygons (output from the rasterization phase) with colors and compute the antialiasing[1] if needed. Each fragment is associated with a color. Usually triangles are used as polygon output from the rasterization phase, since they have simple shapes and computations are faster (Figure 1.5).

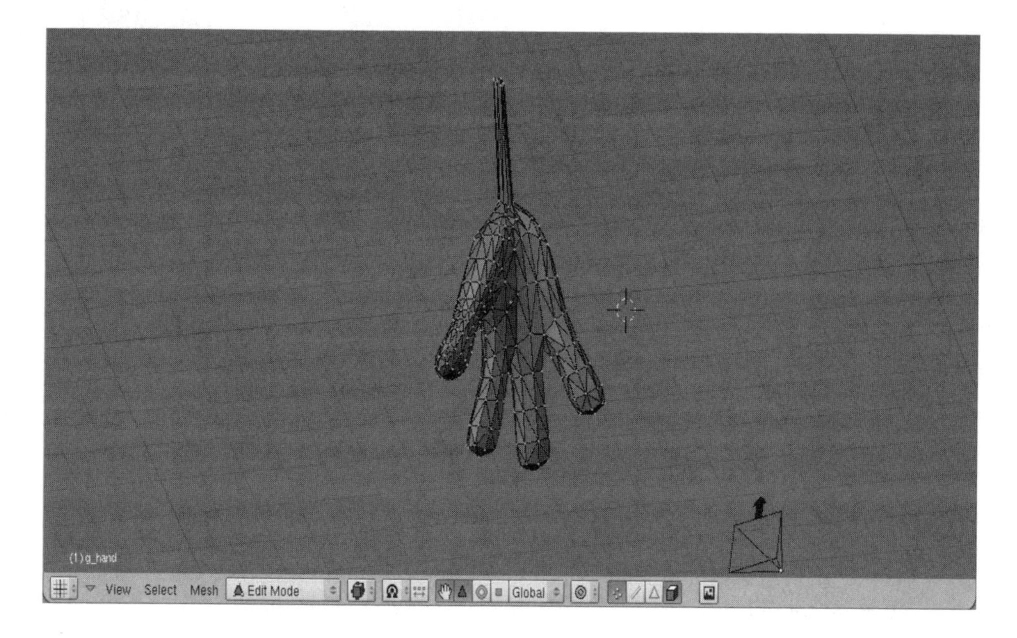

Fig. 1.5. Usually triangles are used as polygons output from the rasterization phase. The object model of the hand is downloaded from [44].

Before writing fragments into the buffer, textures will be considered. Textures are images (like a bit map that could represent a material or a logo) that can be copied onto pixels for obtaining a more photo-realistic rendered scene (Figure 1.6).

After considering textures, the frame buffer is filled with fragments, and finally it is copied into the video memory to produce the 3D scene.

We can subdivide a 3D scene rendering in many levels of photo-realism:

- Orthogonal wire-frame rendering
- Perspective wire-frame rendering

[1] Antialiasing is a smoothing effect; it makes everything look less jagged.

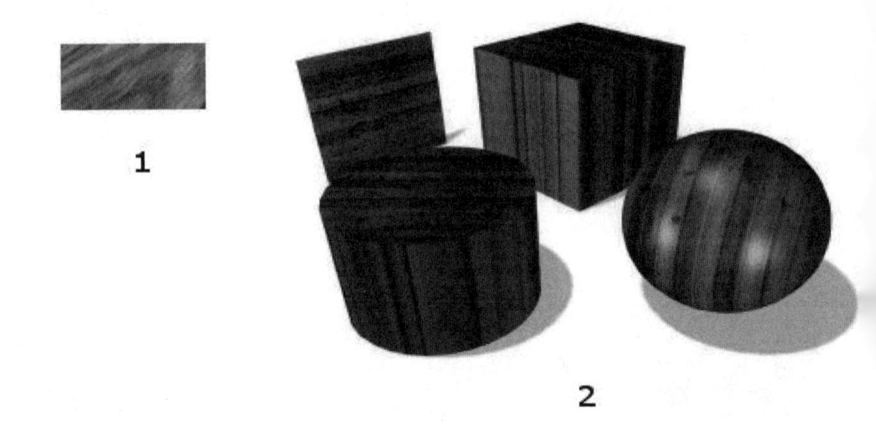

Fig. 1.6. 1: A bit map image for the wood. 2: 3D objects with wood textures and shading.

- Depth cuing and hidden surfaces
- Lighting models
- Textures

1.3.3 Wire Frame.

The first rendering level is called *wire frame*, in which a three-dimensional shape is described by a geometric model made of lines. In the basic version, wire-frame rendering consists of representing all vertexes included in the geometric model; it's mainly used during interactive geometric modeling since it doesn't require many computational resources. Representing all vertexes, the system displays lines, which practically aren't visible, because they are occluded by other opaque surfaces.

This kind of representation is also very useful for technical drawings, in particular for managing orthogonal projections. The main limit of the wire-frame technique is in representing the scene depth effectively, and thus it's not clear which polygon faces are visible and which are invisible faces of a surface, as shown in Figure 1.7.

During modeling of three-dimensional scene surfaces, it is important to be able to change parameters referring to a certain surface in order to shape vertex and surface orientation by using geometric normals.

1.3.4 Depth Cuing

The *depth cuing* technique is frequently used when the computational costs for removing invisible lines (occluded by other surfaces) from the scene are

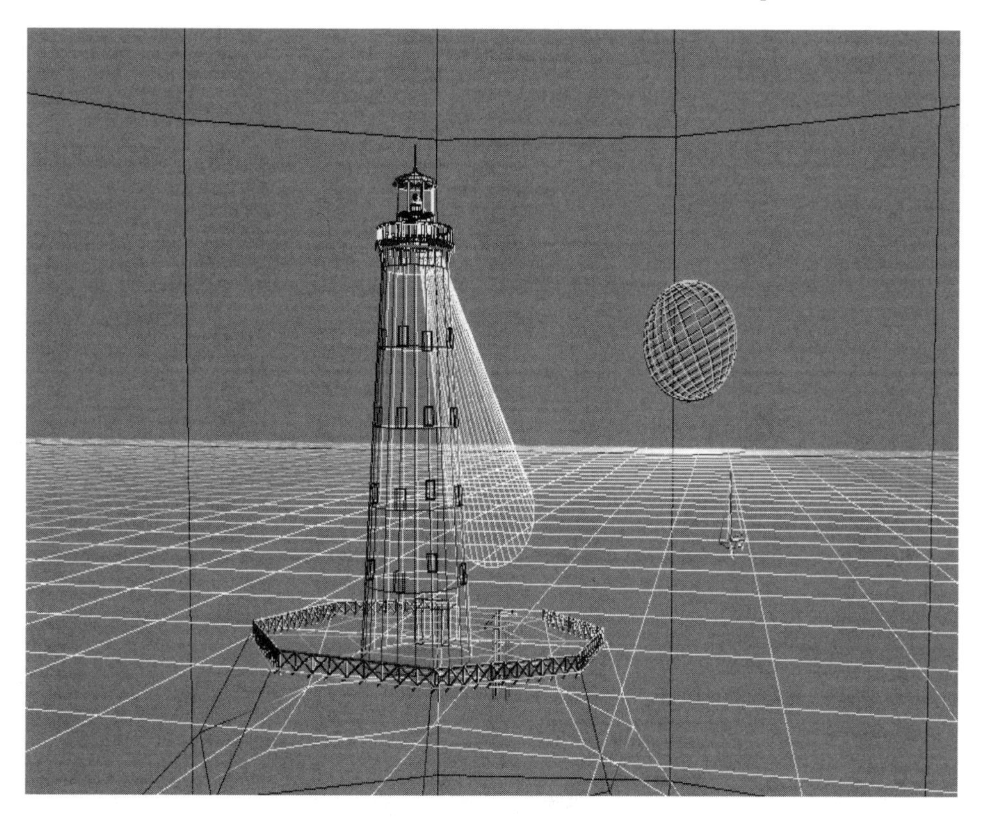

Fig. 1.7. A wire-frame representation.

high. It consists of assigning to every segment a gray color whose saturation increases with depth in the scene. This technique is inspired by the perspective technique used by Leonardo Da Vinci, and helps in understanding proportions among different parts of a complex scene. The solution consists usually in giving high brightness levels to segments close to the front plane (with respect to the observer) and low brightness to those close to the back plane; all the other segments have intermediate values according to their depth in the scene, as shown in Figure 1.8.

1.3.5 Hidden Surfaces

Hidden surfaces have been a severe problem for a long time, especially for their computational complexity; today it is manly solved by using dedicated graphics processors and the *z-buffering* technique.

To solve this problem, many approaches have been researched. The hidden surface removal problem could employ many solutions that involve different efficiency levels depending on the *consistency* of a scene. The concept of *consis-*

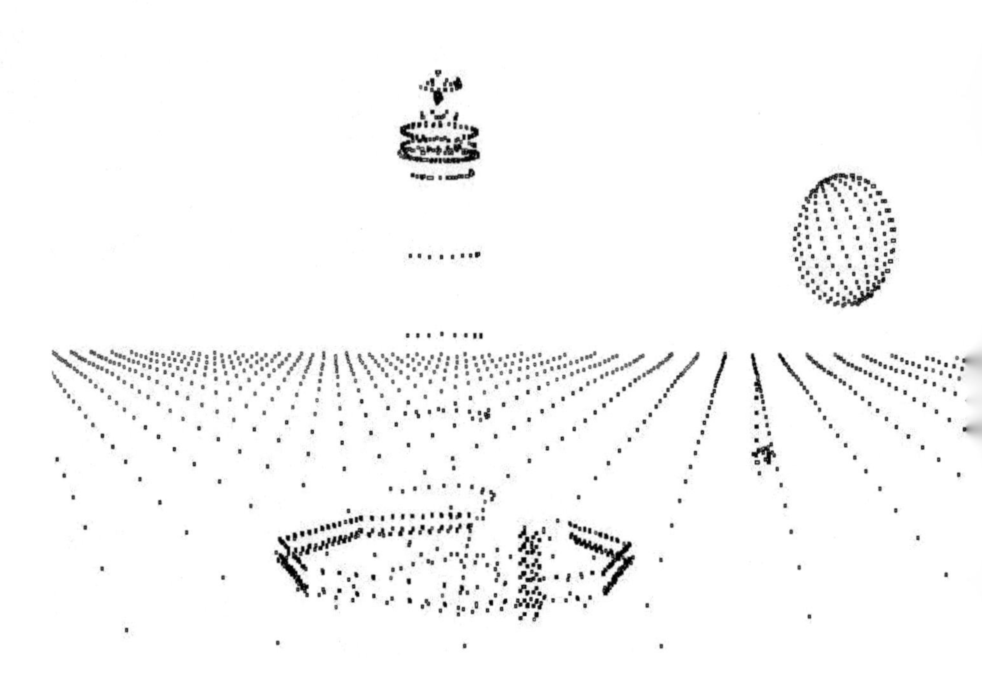

Fig. 1.8. Depth cuing effect on a lighthouse model.

tency is very interesting, and it's usually expressed in many ways in computer graphics. It consists of matching individual and relative surface positions and orientation.

There are eight kinds of scene consistency in three-dimensional modeling that could be used for solving the hidden surface removal issue [10] (Figures 1.9). To clearly understand these kinds of consistencies, we recall that the scene is evaluated after perspective transformations, just before planar projections, and thus depth information is preserved.

The consistency properties are as follows:

- *Consistency among objects*: if two objects are spatially split, then each object face is spatially split (for hidden surfaces, if two objects don't overlap each other, their respective faces don't overlap).
- *Consistency among faces*: generally properties of a face change gradually, from one position to another (if two faces overlap each other in one point, it's likely that overlapping will be spread all over the face; this could be tested very quickly).

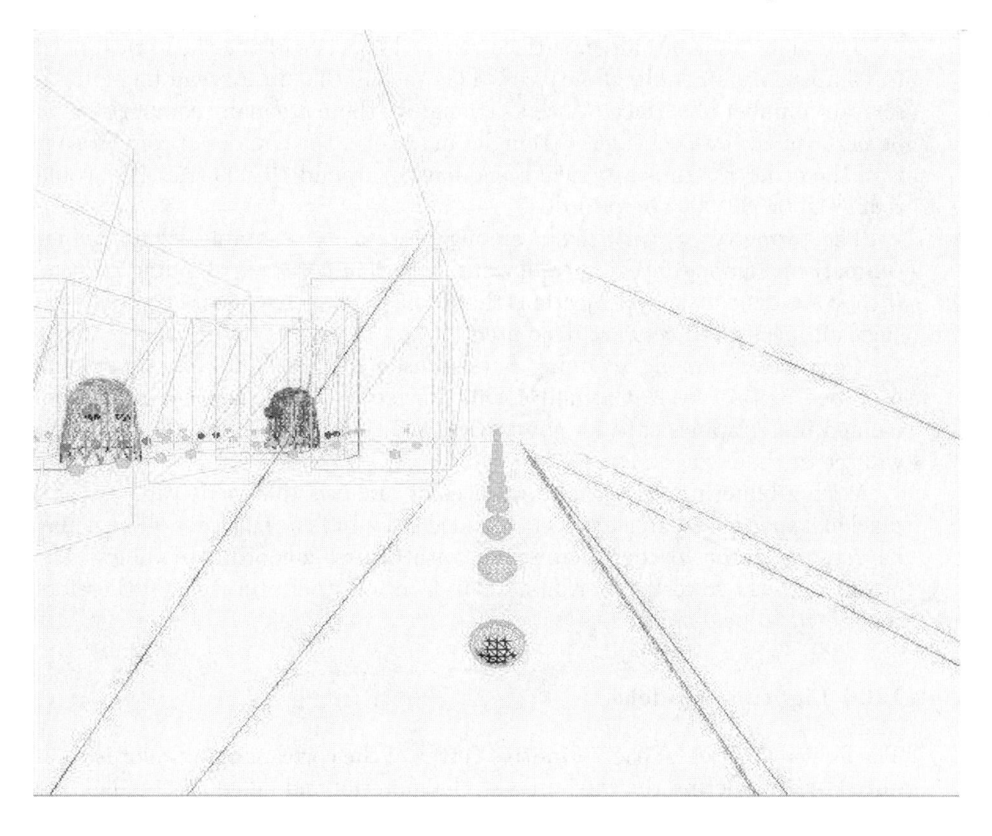

Fig. 1.9. An example of a scene without hidden surface removal.

- *Consistency among vertexes*: the visibility condition of a vertex changes only if the vertex perforates a face or if it intersects another visible vertex.
- *Consistency among intersecting vertexes*: if one face perforates another one, their intersecting vertex could be computed by two intersecting points (thus the face intersecting computation is faster).
- *Consistency among scanning lines*: visible objects along a scanning line are likely to remain in the subsequent scanning line.
- *Area consistency*: a group of pixels that are close to each other is usually covered by the same visible face.
- *Depth consistency*: portions of the same faces close to each other are also close in depth, while different surfaces positioned over a projection plane are usually distinguished in depth.
- *Consistency among frames*: two projections of the same scene computed from two different viewpoints in a sequence of frames are likely not to differ so much; in fact, computations made for a frame could be partially reused for subsequent frames.

The main limit of hidden surface removal consists of the theoretical need to compare each couple of surfaces. This means that in a scene including a relevant number of surfaces (such as triangles), there are many comparisons to be performed. If we consider n triangles in a scene, the number of comparisons is of the order n^2; thus a typical scene having around $10,000$ triangles would require $100,000,000$ comparisons.

The property of *consistency* among objects, for example, could reduce comparisons among faces by comparing *bounding boxes* surrounding objects. If they are disjointed in projection, there's no need to have more comparisons since all faces of objects are disjointed.

Consistency among scanning lines is used in the *z-buffering* algorithm, proposed in 1974 by E. Catmull; for its efficiency, it has been directly implemented in a graphics card for supporting real-time 3D scene rendering, as we will see in the next section.

With z-buffering, the graphics processor archives the z-axis value of each pixel in a specialized area of memory called the z-buffer . Different objects may have the same (x, y) coordinates, but with diverse z-coordinate values. The object with the lowest z-coordinate is in front of other objects, and therefore is selected to be displayed.

1.3.6 Lighting Models

The lighting model for each primitive vertex of the corresponding color is computed by considering the attributes of the material and lights. Vertex lighting information requires defining the geometric normals in order to manage the object reflections for the selected vertex.

Usually there are four lighting types [11]:

1. **Ambient**: light that comes from all directions equally and is scattered in all directions equally by the displayed objects. It's a first approximation for light that comes fairly uniformly from the world and arrives onto a surface by bouncing off so many other surfaces that it might as well be uniform.
2. **Diffuse**: light that comes from a particular point source (like the sun) and hits surfaces with an intensity that depends on whether they face toward the light or away from it. However, once the light radiates from the surface, it does so equally in all directions. It is diffuse lighting that best defines the shape of 3D objects.
3. **Specular**: as with diffuse lighting, the light comes from a point source, but with specular lighting it is reflected more in the manner of a mirror where most of the light bounces off in a particular direction defined by the surface shape. Specular lighting is what produces the shiny highlights and helps us to distinguish between flat, dull surfaces such as plaster and shiny surfaces like polished plastics and metals.

Fig. 1.10. A wire-frame scene from a 3D version of the PACMAN game.

4. **Emission**: in this case, the light is actually emitted by the polygon equally in all directions.

Figure 1.10 and 1.11 show not only z-buffering but also lighting effects, solid colors and Figure 1.12 and 1.13.

1.3.7 Textures

Texture mapping is an approach for adding realism to a computer-generated graphic. An image (the texture) is mapped to a geometric silhouette that is created in the scene, so it is glued to a flat surface.

The resulting pixels on the screen are calculated from the texels (pixels of a texture), and is managed by texture filtering. The fastest method is to use exactly one texel for every pixel, but more complex techniques exist.

Fig. 1.11. The same scene as Figure 1.10 (with a slightly different viewpoint), after applying z-buffering.

1.3.8 Mobile Graphics Software

There are several standards emerging in the mobile market today. Mainly these come in the form of API, that can be integrated using different kinds of programming languages like C, C++, or Java. To develop graphics applications and games, programming skills are required, but they aren't the only thing these kinds of applications need; content and graphics designers are also required. Thus newer standards can be considered as a framework integrating imaging, animation, and geometry representations with programming API.

OpenGL is a well-known platform for people working in 3D graphics and games; graphics workstations used to have video cards supporting its API. In recent years the OpenGL library has grown, with a lot of new API, including API that are rarely used by programmers. There is also plenty of reference material available in the form of books, tutorials, and "how-to" documents downloadable from the Internet.

Fig. 1.12. A lighthouse model with solid colors.

Recently there has been much discussion about 3D graphics in the emerging mobile market. It was unclear how to provide a good user experience using graphics and games on handhelds with relatively small screens and limited processing power. These issues generated uncertainty in the graphics and games developer communities for mobile device manufacturers who wanted to start integrating 3D features into their applications.

To address these issues, a consortium of companies called Khronos developed a version of OpenGL for embedded systems, the so-called OpenGL ES. OpenGL ES is a standard API set for advanced 2D and 3D graphics on handhelds and mobile devices, providing graphics interfaces between hardware and software [12, 13]. Both programmers and content creators need tools for design and development, using state-of-the-art techniques on mobile devices. OpenGL ES satisfies these requirements, allowing them to create their best applications and games incorporating what I call the "mobile gravity rule." This rule describes interaction among objects, where the force is the inverse of the distance. In our case the inverse is in between the screen size and the software rendering capabilities. The smaller the screen, the higher is the pro-

Fig. 1.13. The same scene as Figure 1.12 after applying texture mapping.

cessing power per pixel needed to maintain a good graphic rendering of the scene. Since the Khronos group was started, the OpenGL ES standard has received wide support from over 50 companies including Nokia, Ericsson, Motorola, Qualcomm, Sun Microsystems, as well as the Tao Group, Symbian, Fathammer, Superscape, and Vicarious Visions. These companies are working together to make OpenGL ES a royalty-free open standard for the mobile 3D graphics applications and to show the advantages of OpenGL ES to the mobile developer community.

The OpenGL ES specification is freely available for download, so anyone can develop applications based on it, royalty-free 1.14. The goal of OpenGL ES is to take into account the real capabilities of mobile devices such as no dedicated floating-point hardware (although ARM vector floating-point coprocessor-enabled devices could also support floating point) and lack of memory. It includes the minimum set of API needed for development of mobile graphics. Since OpenGL ES has low level API, programmers can get close to the hardware for high performance applications such as interactive software and games.

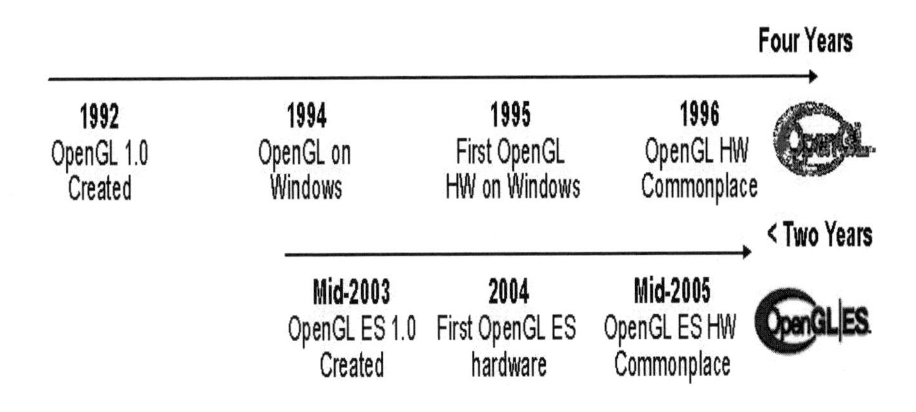

Fig. 1.14. OpenGL and OpenGL ES roadmap.

The OpenGL ES API defines a graphics processing pipeline (Figure 1.15), supporting individual calls to be executed on dedicated hardware, run as routines on the system CPU, or implemented as a combination of both dedicated hardware and software routines. This means that software developers can use a software 3D engine today, and seamlessly transition to using OpenGL ES hardware-acceleration on future devices. Moreover, OpenGL ES enables new hardware innovations to be accessible through the API via the OpenGL extension mechanism. As extensions become widely accepted, they are considered for inclusion into the core OpenGL ES standard.

From a developer point of view, in order to use these API the mobile device OS must be open and allow users to install new applications. Usually mobile devices, and especially mobile phones, are closed in this sense, but things have started to change. Open platforms such as Symbian OS and Java ME (Micro Edition) are counterexamples and adopted standards. Jumping ahead, developers could not only develop using OpenGL ES API but also support the specification JSR-184 as a 3D API for J2ME (Java 2 micro edition, Figure 1.16).

The OpenGL ES standard API is at a lower level than the J2ME extension, JSR-184 API for mobile 3D graphics. The two different technologies integrate into the mobile graphics pipeline at different levels, and thus they can be combined. This book is mostly about how to use 3D graphics API for developing cutting-edge 3D graphics applications without losing good graphics quality. You will learn about developing computer graphics applications for mobile devices using examples taken from different API. We will concentrate on researching how to simulate or enable quality 3D features that we usually find on our desktop applications.

In fact, OpenGL ES is fundamental for standardized access to hardware acceleration solutions for mobile devices. However, coding all the graphics data into a low-level application can result in a huge file requiring too much

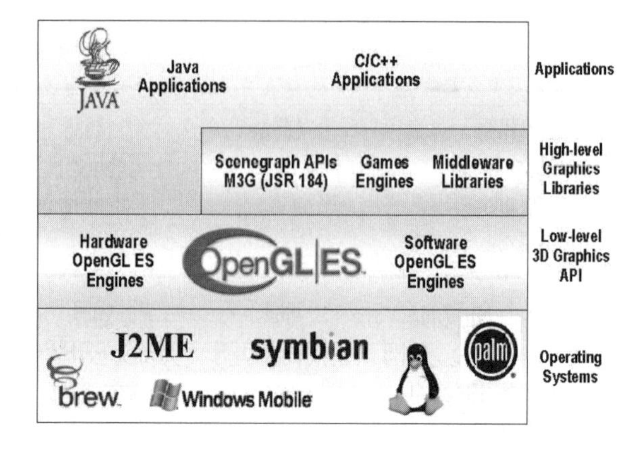

Fig. 1.15. Mobile 3D graphics pipeline layers.

memory for a mobile device or too much bandwidth needed for transferring among devices. JSR-184 helps in this case because it allows graphics designers and developers to define a scene with a platform-independent set of API (Java-based), simplifying the production and distribution of contents. JSR-184 stands on top of OpenGL ES API, and a device supporting both standards will benefit both from hardware acceleration and an abstraction layer. Moreover, both API enable applications to run on products ranging from mobile phones to workstations, making it easy and affordable to offer a variety of advanced 3D graphics and games across all major mobile and embedded platforms [13].

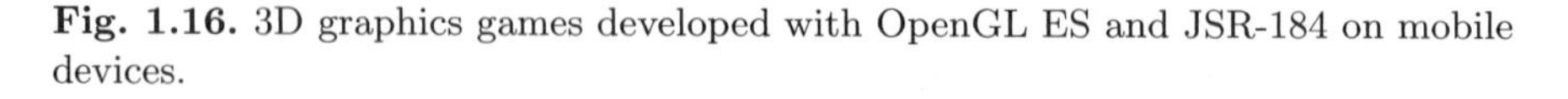

Fig. 1.16. 3D graphics games developed with OpenGL ES and JSR-184 on mobile devices.

One question to address is, considering the resource-limited handhelds, how can they efficiently run Java programs, as they tend to be slow even on graphic workstations? In fact, usually Java programs require more processor overhead in order to run. On the other hand, there is a need for features like a scene graph representing the geometric structures, hierarchies, and camera points of view of objects in a 3D scene. JSR-184 is the answer; it is a higher level set of API and thus programs written on top of this specification can be reused on many different devices. The JSR-184 is complementary to OpenGL ES, and the rendering modes are compatible, so the graphics hardware that accelerates OpenGL ES will also accelerate JSR-184 API. Three-dimansional graphics on mobile devices are rapidly growing in response to market demand. JSR 184 is already a requirement for major operators worldwide, and devices that implement the API are already on the market.

In the following chapters we discuss how researchers and developers can build 3D graphics applications that perform well on mobile devices, resolving issues concerning limitations in display size, number of colors, processing power, and the power consumption of handheld devices.

In particular, some of the subjects we discuss include setting up a graphics window; managing color, rotation, and translations; building 3D shapes, texture mapping, filtering, and lighting; blending, loading and moving in 3D space; display lists, bit-map fonts, 2D texture fonts, fogging models, quadrics, particle engines, lines, and antialiasing; bump-mapping and multi-texturing; morphing and loading objects from a file; clipping, reflections, and shadowing.

1.4 Summary

In this introductory chapter, we have surveyed the major hardware and software features of mobile and handheld devices. Different kinds of hardware, such as handhelds, smart phones and mobile phones, have been discussed. The three major operating systems have been presented in order to have a clear idea of the platform strengths and limitations on top of which we will build our mobile 3D graphics applications. We discussed standard graphics software packages and presented a model for the coordinate system pipeline. Graphics programming packages require coordinate specifications to be given in a Cartesian reference frame. We discussed the many different changes in the coordinate systems from the model to the rendering of the final 3D scene. We explained that functions available in graphics programming packages can be divided into graphics output primitives, attributes, geometric and modeling transformations, viewing transformations, input functions, and control operations. Concerning software packages, we described the OpenGL library consisting of a device-independent set of routines for managing 3D graphics. We then focused on mobile 3D graphics presenting the main issues, both from the performance and the appearance points of view, considering the limitations of handheld devices. We described the main software packages that will be presented in this book, the OpenGL-ES and JSR-184 specifications. These packages work well together, providing both low- and high-level access to mobile graphics features. We will explore how to obtain graphics results on mobile devices as good as those on workstations.

2

Mobile 3D Graphics: Scenarios and Challenges

2.1 Application Scenarios

Many application fields can benefit from mobile 3D graphics, and some that are already adopting the technology include, mobile tourist guides, augmented reality, and mobile gaming.

In the last few years, tourist guides on mobile devices, also called *mobile guides*, have aroused a lot of interest in the research field, both in industry and academia. The main advantage of mobile guides versus traditional guides consists in offering updated services using handheld devices, that are easy to carry, like the PDA or new-generation mobile phones. We can identify two kinds of services provided by a mobile guide:

- *Navigation support*: provides users with driving directions.
- *Information delivery*: provides users information on points of interest, like banks, gasoline stations, hospitals, etc. in the user's vicinity.

Until now, all the graphics representations in these services were made of 2D maps with text labels. With the introduction of 3D graphics capabilities in mobile devices, many researchers found that using 3D modeling with the mobile guides allows user interfaces to be more intuitive and friendly for these kind of applications. Many usability studies have shown that 3D environments help in understanding real spatial relationships, and thus could help users in navigating both the interfaces and the environment. In fact, by using a 3D representation, the viewer's perception of the environment improves, since users can better understand distances and proportions between objects visualized by the mobile guide and the real environment.

An example is the prototype of a 3D city, built by Vainio and Kotala [14], using a three-dimensional city model, a map, and a database. The scope of the prototype is to help users in navigating the environment, showing information on surrounding geographic space by a real-time link between the 2D map and the 3D visualization, as shown in Figure 2.1. The prototype running on a PDA allows users to see the screen split in two different parts: in the upper part there's the 2D map, and the lower part displays the 3D model of the surrounding area. Usability tests provided with this study show how users prefer to manage the 3D model combined with the 2D map rather than choosing one single view, either the 2D map or the 3D model. Moreover, users can clearly identify their position and orientation using the 3D virtual model rather than the 2D map.

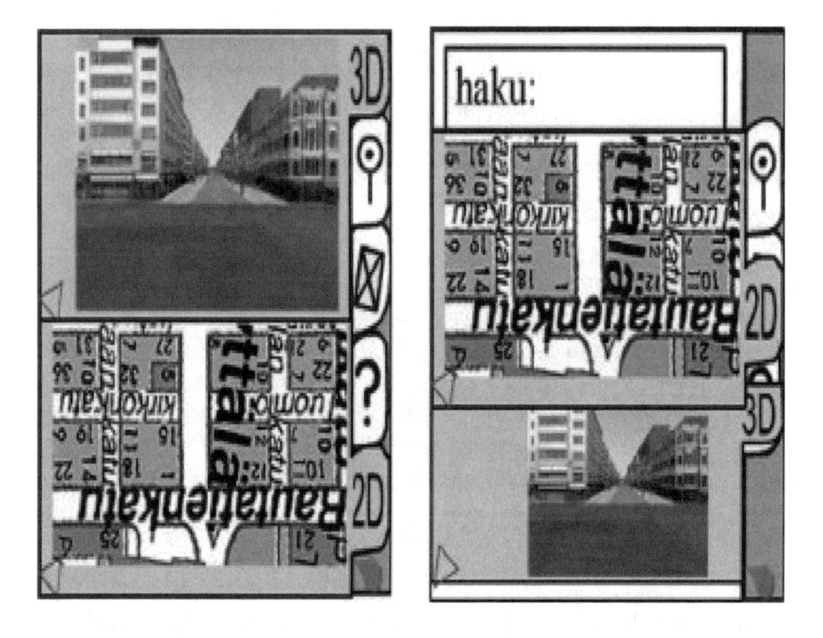

Fig. 2.1. The user interface for the 3D City Info research project.

The research group at the HCI Lab of Udine [15], Italy, has developed another mobile guide prototype, which is context sensitive and uses 3D graphics for rendering the user's surrounding areas. This prototype allows users to interact with virtual displayed objects by selecting them directly and obtaining textual information.

It's interesting that in this prototype we can identify three main navigation modalities:

- *Manual navigation*: the user can manually change the point of view in the 3D virtual environment.
- *Automatic navigation*: a global positioning system (GPS) is used in order to react in real time to the new user position and display it in the virtual environment.

- *Replayed navigation*: the system uses prerecorded positions and orientations or paths designed by a human guide, or by the user in order to display a virtual tour in the environment.

It's clear how 3D views are relevant, especially for the latter two modalities, where the user could choose to have a snapshot of the areas of interest before reaching them, and plan the visit or make virtual tours seated at home in his favorite chair.

The future direction of mobile guides will involve all the algorithms and techniques that we will describe in this book, both at visualization and interaction levels.

In fact, as researchers suggested, in some situations the user might want to see a map larger than the city model and thus the interface should be adaptable to meet his needs. In other cases the user might want a three-dimensional model, instead of a map, in order to find a specific place like a fountain or an ancient front door. An example could be the *Google Earth* application [16], where users, by means of zooming, can start from the visualization of the whole earth and navigate, finally reaching a house. Thinking at this level, if the user can zoom more deeply until reaching a fountain 3D model, it would really be useful. This is a clear example of how integrating zooming with different resolution graphical levels could increase and enhance the user's experience. It turns out that with this approach the user could directly select the objects of interest in the 3D representation. In fact, as suggested by previous research, the easiest way for the user to ask for information about a building or some other point of interest is to point at it with the finger. The use of a stylus could replace the finger in the virtual model representation.

Another desired feature relates to proximity. When a user gets closer to a building, the point of view and the perspective must change following the natural inclination of the head. In fact, getting closer to buildings or churches we tend to look upward to the roof and change the perspective while moving in order to have a complete view of the subject. This again helps in developing a 3D effective visualization of the surrounding scene in a mobile guide, and with the technology provided by new mobile graphic libraries we can achieve results comparable to those done on a desktop computer.

Augmented reality (AR) is a research field that aims to combine real-world information with computer-generated data to enhance perception of the surrounding environment. The main idea involved with augmented reality is about enriching the user's experience with computer-generated information. Usually this happens using special head-mounted displays (HMDs).

Initially, AR was introduced apart from virtual reality; instead of immersing a user in a virtual environment, the goal of augmented reality is to enhance the real world with information not available in reality. Someone suggested that virtual reality is a special case of augmented reality since the AR adds to the real-world virtual information that virtual reality reproduces. A simple example of AR is a football game on TV. The real-world elements are football players and the field, while virtual elements are the lines displayed during the replays to highlight penalty analyses, such as players' being offside, or circles displayed to show the right distance to maintain for the penalty kick or the partial score displayed on the field.

Milgram and Kishino [17] describe a taxonomy that identifies how AR and virtual reality are related. They define the reality-virtuality continuum shown as Figure 2.2.

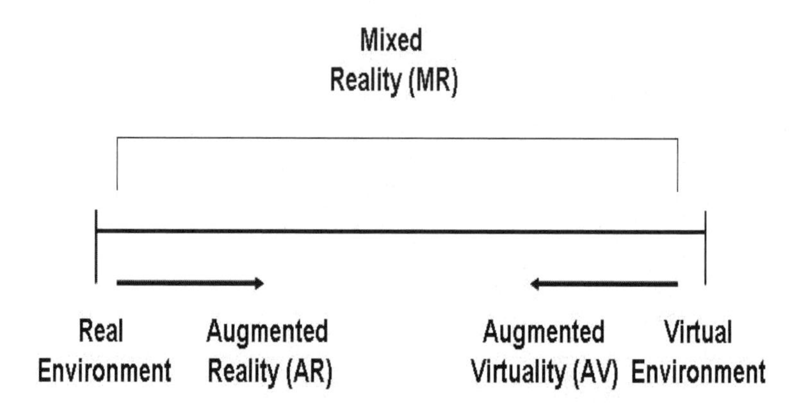

Fig. 2.2. Milgram's reality-virtuality continuum.

The real world and a virtual environment are at the two ends of this time-space extension, with the middle region called mixed reality. Augmented reality is placed near the real-world end of the line, with the main perception being the real world augmented by computer-generated data. *Augmented virtuality* is a term created by Milgram to classify systems that are generally synthetic with some real-world graphics added such as texture-mapping video onto virtual objects. This is a differentiation that will disappear as the technology improves and the virtual objects in the scene become less detectable from the real ones. Milgram further describes a taxonomy for the mixed-reality displays. The three axes he advocates for grouping these systems are Reproduction Fidelity, Extent of Presence metaphor, and Extent of World Knowledge. Reproduction Fidelity describes the quality of the computer generated imagery ranking from simple wire-frame approximations to entire photo-realistic renderings. The real-time requirement on augmented reality systems forces them to be almost the low end on the Reproduction Fidelity spectrum. The current graphics hardware power cannot produce real-time photo-realistic renderings of the virtual scene. Milgram also assigns augmented reality systems on the low end of the Extent of Presence metaphor. This axis assesses the level of involvement of the user within the displayed scene. This classification is closely related to the display technology managed by the system. There are considerable classes of displays employed in AR systems. Each of these contributes to a different sense of immersion in the display. In an AR system, this situation can misdirect because with some display technologies part of it is the user's explicit view of the real world. Immersion in that display arises simply by looking at the scene. It is differentiated to systems where the combined view is displayed to the user on a separate monitor for what is sometimes called a "Window on the World" view. The third, and final, dimension that Milgram adopts to categorize mixed-reality displays is Extent of World Knowledge. Augmented reality does not simply denote the superimposition of a graphic object straight onto a real-world scene. This is technically an easy task. One problem in augmenting reality, as described here, is the need to conserve precise registration of the virtual objects with the real-world image. As will be explained, this often includes accurate knowledge of the relationship between the frames of reference for the real world, the camera viewing it, and the user.

We can figure, as we will see in this discussion, that AR applications with mobile devices will be placed at the high end of the Extent of World Knowledge metaphor.

Today, many systems and applications have been studied for AR. Usually they include the use of a single personal computer, usually a notebook, connected to external environment sensors and devices like custom display-based glasses that show the real environment plus some virtual information.

The existing AR systems need a cumbersome hardware infrastructure, thus limiting the arm and body movements in the real world. Another problem is that this kind of system is very expensive, and this is a limit in developing these technologies. Indeed there exist situations or social environments where these expensive and awkward technologies are inadequate. The best solution will be to use lightweight devices and thin clients[1] easy that are to carry or wear with an adequate network infrastructure.

The mobile devices and capacity-enhanced graphics are accelerating the development of augmented reality systems on personal digital assistant (PDAs) and smart phones. From a mobility point of view, a PDA is easier to carry than a notebook and doesn't limit movements since it is light enough to be carried in the hand.

One of the first institutions to study so-called *handheld augmented reality* was the Vienna University of Technology, with the development of an application for 3D navigation in a predetermined environment. Daniel Wagner and Dieter Schmalstieg [18] described the first stand-alone AR system with self-tracking running on an unmodified PDA with a commercial camera. The project exploited the ready availability of consumer devices with a minimal need for infrastructure. The application guides a user through an unknown building by showing a variety of navigation hints, including a wire-frame visu-

[1] A thin client is a computer (client) in client-server architecture networks which has little or no application logic, so it has to depend primarily on the central server for processing activities.

alization of the building structure superimposed on the video image, together with labeling of relevant elements and highlighting of the next exit to take, as shown in Figure 2.3.

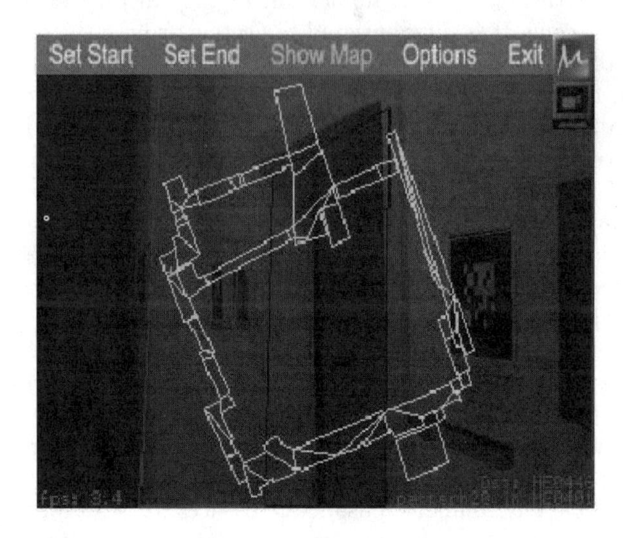

Fig. 2.3. The application called Signpost [18]; showing the environment overview map.

Another interesting example of AR is a game called the Invisible Train [19] (2.4). The game uses several PDAs with displays pointing at a small railroad. The displays serve as a virtual lens showing two small 3D trains that users can drive using the touch-sensitive display. They can increase or decrease speed, and avoid collisions. The players' PDAs are synchronized via a wireless network.

All these studies and implementation have been developed on top of custom framework software that has performance problems with 3D visualization. This is due to the fact that at the time when these projects were being developed there weren't standard and efficient graphics libraries for 3D graphics on mobile devices. And again we will see how using these new standard approaches and technologies will help in developing AR applications and take advantage of the light and compact nature of handheld devices, without major limitations.

The *mobile gaming* and more generally the mobile entertainment field are contributing to and pushing for development and enhancements of mobile hardware and software platforms. Many software companies and video card manufacturers have invested a remarkable amount of money in developing interactive and high-end graphics applications to try to obtain the same good results as those achieved for desktop computers and gaming consoles, dealing with problems that will be discussed later in this chapter.

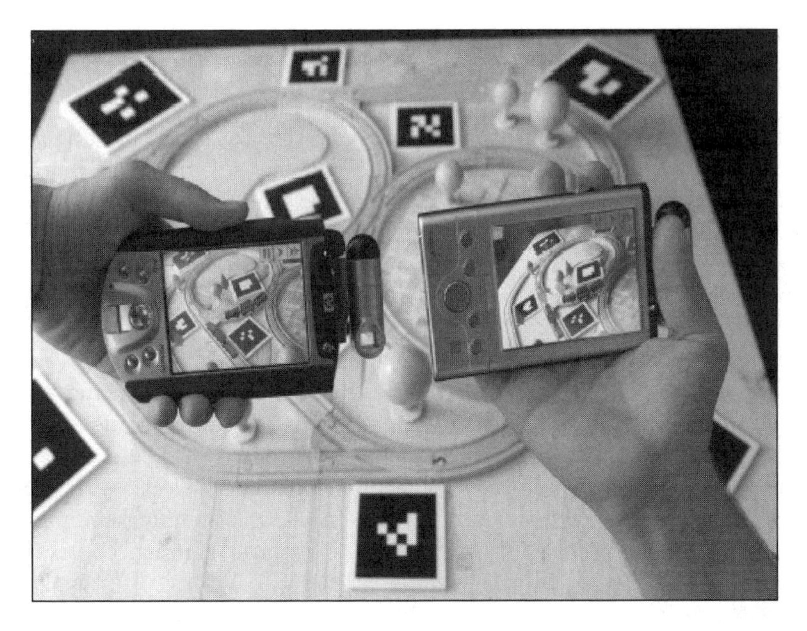

Fig. 2.4. Two PDAs running the Invisible Train game [19].

Moreover, many market studies have predicted that the earnings, at a worldwide level, for mobile video games will rise from $500 million in 2002 to $2 billion in 2006. In fact, if we look at Tetris, a simple two-dimensional mobile game by Blue Lava, we can see that it obtained one million paid downloads in less than one year.

One of the main aspects of this market and the reason for the huge amount of investment is the wide potential audience for games. Many companies have estimated that the annual purchase of mobile devices and PDAs will be between five and six millions of units yearly. The desktop computer market is about two million units per year, which is why there is so much interest and investment in mobile devices and mobile gaming and entertainment particularly.

The Game Developer Association [20] has recently released a white paper on the status of the mobile gaming industry describing the issues concerning these new technologies as well as the advantages and benefits that the mobile technology could lead to. Mobile games design and development, while having many similarities with the computer games industry, also has many issues to be addressed that are different and new. The design and development of mobile games has to deal mainly with the product life cycle, which is short, from 4 to 12 months, more or less half that of the consoles or desktop PCs. This is similar to mobile phones market, where the last product on the market becomes obsolete in few months.

Other issues are related to the variety of mobile devices on the market, with each supporting different standards. Moreover, we need to think about the fact that every device could be used in a different worldwide location, so developers have to think about localization (languages, interfaces, ...) and packing all this information in a resource-limited device. There are advantageous marketing possibilities: people can buy a game to be played once, or they can take a pay-per-play subscription and even pay with micro-payments, like using Short Message Service (SMS), for buying video games. So the distribution is very cheap and simple, because a game could be easily downloaded from the mobile network.

The vast majority of games today utilize 2D graphics, have no or very little multiuser capability, and are very simple. However, this is changing rapidly, and there are now a significant number of ground-breaking games that employ 3D graphics, multiuser features such as turn-based play, and community features such as shared high-score tables and opponent selection. The possibilities that a mobile device could provide in terms of gaming compared to a console are wider, especially considering multiplayer gaming. For example, Nokia has recently released a new game titled HinterWars [21]. It is the first multiplatform PC and mobile game that allows people to play with millions of other people, both from their home desktop PCs or alternatively with the mobile device Nokia N-Gage. For the first time ever, a game uses the real mobility concept; it can be played in every location: from home with your own PC, but also on a bus or, taxi, or at the office via the network.

Since mobile devices are network-based devices, it seems obvious that multiplayer gaming would expand into the Internet world. There are still some problems to overcome in mobile multiplayer development, like power consumption and latency. These issues have to be considered during the mobile games design phases in order to provide users with real-time feedback to their actions. Even if new standard networks are developing, like 3G, they still aren't supported worldwide, so for now designers have to manage network latency. Many games now offer a centralized high-score table to compare results from users in different locations, thus enabling players to compare their results, therefore stimulating the competition. The challenge is to be able to create first-person shooters, racing games, and real-time feedback games where high frame rates are a key to achieving high throughput of data.

An interesting variation of multiplayer games are community games. Teams or groups of users will play together for the goal of the game, sharing their playing experience fighting or being allied for a common aim. Technologies like SMS and global position system (GPS) could be used in conjunction to provide players a whole new set of experiences, like playing virtual games in the real world.

Community players often, if not always, connect and link to others in pursuing a game target together; this highlights the entertainment capabilities of mobile devices and raises the demand for including 3D graphics support on the mobile devices.

The actual mobile devices are not so ready to deal with three-dimensional entertainment applications. There are still many issues related to the application's usability, such as input devices that are uncomfortable for their small size, especially for dealing with 3D graphics interactions. Also the performance isn't capable of dealing with fast 3D scene rendering; they're slow compared to medium-resolution desktop PCs. The main enhancements should be provided both at the design and the hardware level to fill this gap.

Usability is fundamental for PDAs [22] and mobile devices. For example, consider integrating e-mails, instant messaging, online navigation, and voice telephony all in single device. There are already devices, on the market capable of supporting all these features, but many times they're not very usable. For instance, screens are still too small to browse the Internet or read long e-mails without the need for continuous scrolling. Wider screens will work around these issues but designers will probably have to deal with power consumption. For many input tasks it seems that QWERTY keyboards enhance usability, like writing emails or dialing a phone number; also for editing it still seems a preferred solution compared to the stylus with a "graffiti" language to learn (which also has to manage hand-writing noise and errors). New approaches for full navigation on the Web are preferred to support wireless application protocol (WAP) technologies; in fact you can see more information on the screen and go to the next page easily, but pages are still slow to download so that exploring a Web site could be very tedious with a mobile device. There is still a need for a better input device; in fact, online Web sites require long scrolling, and even new tilt scrolling devices (devices that sense the user flicking on them) still aren't good at it. The best solution today would be for content providers to summarize their information and shorten their Web pages to facilitate the mobile user's experience. Moreover, there's no particular tight integration between mobile devices and desktop PCs; usually the synchronization of information is related only to some kind of application, while wider support for this is absolutely required. Otherwise there will be a complete disarrangement in the information users keep on their PCs, thus missing the main goal of a mobile device: allowing you to work with your data or playing your favorite game wherever you are. Finally, there's a need for integration among communication features like SMS and e-mail, for instance automatically recovering a phone number of the sender of your last received e-mail. The best integration will probably be a mixture of actions between mobile device designers and content or Web-site managers, such as shortening the published articles, simplifying navigation, and determining the relevant features for mobile devices related to the desired tasks.

There are plenty of new mobile devices on the market that are starting to support 3D graphics technologies.

The ability to support playing games is obviously the core requirement for such games, but 3D graphics with realistic scene rendering clearly enhance the gaming experience.

Java 3D is supported by using M3G, the Mobile 3D Graphics file format. M3G is available on new handsets from Motorola™, Siemens™, Nokia™, and Sony Ericsson™, among others. M3G also includes a good conformance testing suite, so that there is high degree of compatibility between the M3G-enabled mobile devices. Writing 3D games is fairly similar to writing 2D games; there are assets, game artificial intelligence (AI), and a user interface. The difference is in the rendering and management of the scene. These are managed by an M3G engine, and the assets for the games also have 3D models as wells as 2D bit maps. These models, as well as cameras, lights, animations, surface materials, and appearance, are usually authored in a dedicated 3D tool, such as 3D Studio Max, and then exported to the standard M3G file format. The M3G format is also available for public use, so developers can build exporters for existing tool chains. Thus they can port games from other platforms, such as the GameBoy or PlayStation. We will show and develop a simple customized M3G exporter in later chapters.

Another important standard library for mobile 3D Graphics is OpenGL ES standard, and we will see that it is a cutting-edge technology for 3D mobile graphics, not only as a new engine and set of API but also because it can be downloaded and installed over the Internet. For this reason, each game should support its own 3D engine, rather than utilizing the generic 3D engine provided by M3G on Java handhelds. This has some advantages in that each game can have a carefully customized engine to obtain maximum performance; however, there is also downside in that the engine takes up additional space on the handset, restricting further download. Also development costs can be higher, as each engine tends to have its own authoring tool costs, even if it is only a variation on an existing level editor, for example.

The next generation of mobile devices supporting three-dimensional graphics will be based on the API that will be described in this book, as consequence of a standardization process that is coming out from developers and designers.

Moreover, many new mobile graphics processors have been introduced in the market, and many new generation devices are directly supporting 3D graphics.

In conclusion, we can try to propose a design for a mobile device meeting all the requirements that came out from our analysis, integrating and supporting the three explored areas: mobile guides, augmented reality, and games.

The ideal mobile device will need to support at least a standard OS and standard 3D graphics API. Moreover, it should provide features for playing music, movies, and games. It will need to include effective input devices and support for GPS positioning, not only for localization such as mobile guides applications, but also for multiplayer GPS games where the real user location is fused with game rules, providing augmented reality. This device should also support standard wireless network connections and a small video camera for interacting or recording. Finally, basic mobile phone capabilities should be required, especially SMS facilities both for communicating and supporting micropayments for applications downloading.

2.2 Multimedia and Graphics Usability Challenges

The constant demand for multimedia data is growing exponentially due to continuous enhancement of computing power of mobile devices and software programs that manage such data. Multimedia data are now required not only for commercial and marketing purposes, but also as important design data, with which many applications are developed for example, digital movies, interactive videogames, and office automation applications. Such multimedia applications require an extensive amount of resources in terms of computation, communication, and memory. But mobile devices still have some important limitations: low bandwidth, power, CPU, memory, and storage [3].

For instance, to deal with the problem of the small screen, a number of different techniques have been used. One is the use of a proxy system [23] for preprocessing stages of multimedia objects. When considering text data, the data could be reduced, filtered, or summarized [24], and videos or images could be transcoded by managing frame size, bit rate, or frame rate [25]. Data are presented in a card format on mobile devices, usually by means of thumbnails for navigation. To run multimedia content efficiently and effectively, in [26] metadata are used to validate and represent the multimedia content. For instance, if the multimedia content is computationally expensive it could be replaced by a simpler, less demanding version of the same content.

In many application fields, like digital photography, videosurveillance, telemedicine, and many others, a proposed solution to deal with the image processing needs and the limitation of mobile devices consists of so-called wireless imaging. In these applications it is crucial to be able to capture images and videos and process them for extracting information, like raising an alarm if there is an intrusion in a surveillance environment. To solve these problems a distributed computation approach is needed since mobile devices might not always be capable of performing real-time image processing. So the solution consists of sending images or video frames to powerful servers that process the data and send them back to the mobile device for displaying the results.

An interesting example [27] of an application is in disaster recovery scenarios. Consider a scenario of an archaeological disaster recovery. After an earthquake, a team equipped with mobile devices (laptops and PDAs) is sent to the disaster area to evaluate the state of the archaeological sites and precarious buildings. Their goal is to draw a situation map to schedule reconstruction jobs. Before this process starts, the team leader has stored all area details, including a site map, a list of the most important objects at the site, and previous reports and materials. The team members' PDAs let them execute some operations but don't have much computational power. Such operations, possibly supported by particular hardware (for example, digital cameras, General Packet Radio Service connections, computational power for image processing, and main storage), are offered as software services to be coordinated. After visual analysis of a building, team member 1 (using his or her PDA) fills out questionnaires. The team leader analyzes these questionnaires, with the help of specific software, to schedule the next activities. One team member takes pictures of the precarious buildings, whereas another team member is in charge of the image processing of older and recent photos of the site (for example, to initially identify architectural anomalies). In this situation, matching new pictures with previous ones might be useful. So, the PDA with the high-resolution camera and the PDA with the older stored pictures must be connected, as shown in Figure 2.5.

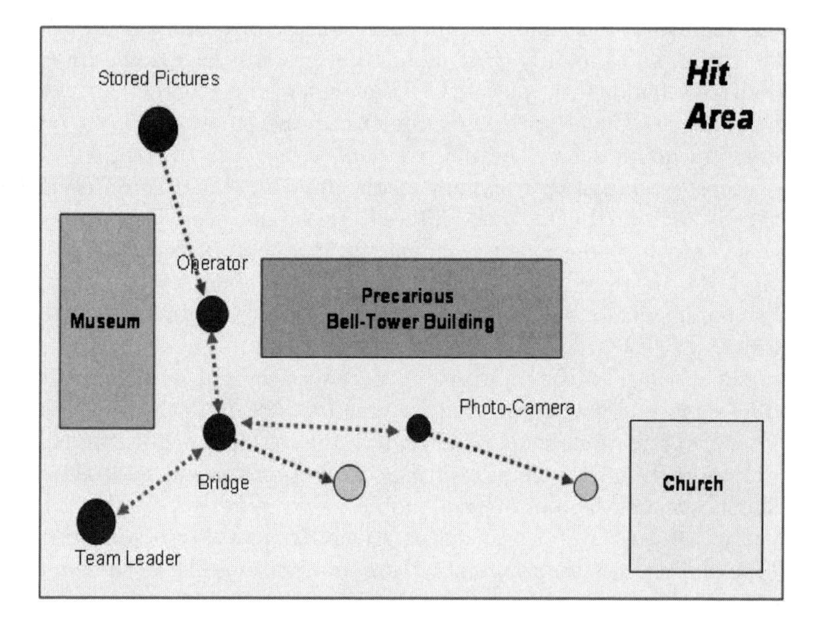

Fig. 2.5. A disaster recovery scenario.

The mobility of handheld devices could be very useful in tasks like face recognition, for instance where a fixed camera is less useful for capturing some expression or using an optimal position compared to lighting and angles. One may think for example of the police using a mobile device for capturing an image of a suspect face while he is on a street, and this would require adapting the image capture according to conditions in the environment. Moreover, to make these kinds of computations in real time, image processing should be optimized in order to be fast enough for capturing and processing information; not only could application compiler optimization be used for enhancing the capturing and processing of multimedia data, but also data could be sent wirelessly to a central or proximity server that could process them and send back results. Alternatively, the server could take some action based on the image processing output.

2.3 Algorithms and Architectural Challenges

In the last few years, many relevant steps have been taken toward design and computation for mobile devices, not only at the technological level but also in research fields. The computational power and communication capabilities have been enormously increased, and also graphics performance and power consumption have been constantly improved. But all these developments still aren't advanced enough to cope efficiently with real-time 3D computer graphics. In fact, mobile devices don't have processors able to process complex mathematical computations; they can't even store a large amount of data (e.g., required for storing textures). Their screens are too small, with low resolution (screens are rapidly improving), and the batteries still tend to run down very fast when dealing with 3D computer graphics (usually this is related to the large amount of computations required).

Among these problems the hardest challenge seems to be the power consumption: mobile devices are following Moore's law [2] both for computational power and for storage or memory capacity, but progresses in battery technology aren't so fast; they increase about 10% in magnitude every year.

Another interesting challenge is the visual quality of image rendering. Actual displays of mobile devices have very limited resolution in graphics rendering. Many common displays have a resolution of 176×144 pixels, while usually the best resolution is of 320×240 (also called QVGA).

The only solution for increasing visual 3D rendering quality and increasing the performance seems to be that of deploying 3D hardware acceleration. For example, let's consider the actual performance: the software implementation for 3D rendering now can manage 1MPixel/s, reaching only a three-frame-per-second performance on a VGA screen by saturating completely the mobile device processor. Currently available graphics hardware could easily perform at 100MPixels/s level.

The advantages of the hardware solution approach are numerous:

- *Better performance*: the processor work load is reduced since the pixel and vertex operations are forwarded to chip specialized for graphics.
- *Better image quality*: advanced graphics functionalities could be implemented during the rasterization phase, such as antialiasing or texture compression.
- *Reduced power consumption*: hardware implementation is less power demanding than software implementation, which involves many other chips in the process.

Hardware acceleration seems to be a good solution to the performance and image quality challenges, but should be carefully implemented in order to fully

[2] The observation was made in 1965 by Gordon Moore, cofounder of Intel, that the number of transistors per square inch on integrated circuits had doubled every year since the integrated circuit was invented.

obtain all the listed advantages. In fact, at the design level one should consider the limited resources available in a mobile device and other issues like overall power consumption and costs; one could easily have a bad design, thus losing the real advantages in supporting this hardware solution. It is thus necessary, for this approach, to have full knowledge of the mobile device architecture in order to identify bottlenecks at the design level and thus avoid wrong design and implementation choices.

The mobile device architecture is very different from desktop computers. It includes only one main memory unit, and the central processing unit (CPU) with the graphic processing unit (GPU) are implemented on the same chip, as shown in Figure 2.6.

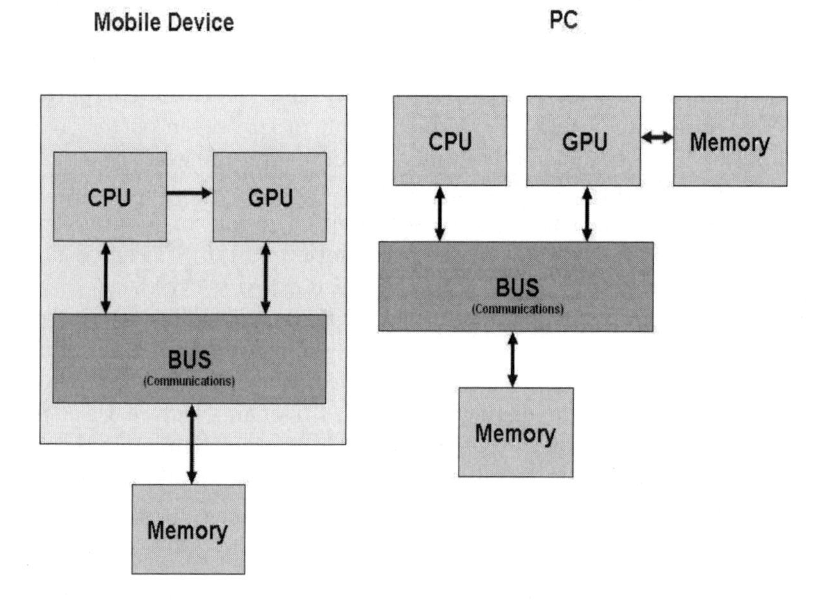

Fig. 2.6. Mobile device VS Desktop PC architecture.

The main issues concerning the mobiledevices architecture are as follows:

1. Limited memory bandwidth. Exchanging data from and to the memory is very time and power consuming.
2. Mobile device processor units don't include a floating point unit (FPU).

The advantages compared to a more traditional architecture are (1) that they are less data exchange demanding since it uses a lower graphics resolution, and (2) that they have an easy and direct access to both the *frame buffer* and *depth buffer* [3] since there is only one main memory.

To improve performance it is necessary to share the computation and tasks among CPU and GPU units, reducing as much as possible the CPU work load. All the computations about lighting and geometrical transformations could be delegated to the graphic hardware, in the same way as for graphics card installed in desktop computers. Usually graphic cards include a programmable unit called *Transform* and *Lighting* (T&L); which is in charge of the two above-mentioned phases. It has to be noted that 3D graphics API usually include profiles. Profiles are configurations that enable or disable certain features depending on the performances and resources available for the target device. For instance, the OpenGL ES includes two kind of profiles:

- *Common-lite*, which provides fixed point computations.
- *Common*, which provides floating point computations.

In general mobile devices don't support a dedicated FPU (in the future they'll probably will), and thus the first profile is more suitable for them. The first profile will also be very useful in producing lightweight mobile devices appliances, and it is a good choice for smart phones and mobile phones in general.

Floating-point operations could be also emulated by software, but using many floating-point operations, if supported by emulation, heavily reduces the performance. Where available, it's better to choose fixed-point math. For example, to reduce pixel data exchange (thus bandwidth), smaller textures could be used, or to reduce pixel transformation, smaller rendering windows are a good solution.

In the per-vertexes[4] operations the goal is to simplify as much as possible the objects, geometry by selecting the same vertex as often as possible. For example, if we consider an approximation of a geometric shape with a triangle, we could use triangle strips, as shown in Figure 2.7 [28], for building n triangles using $n-2$ vertexes. This reduces the number of vertexes and therefore the number of computations. We could imagine that every time a transformation takes place (e.g., rotation) all the vertexes have to be transformed, so the number of vertexes is proportional to the number of cycles required for transformation process.

It is also useful to reduce the information data size for vertexes by choosing the smallest data type possible. For instance, a short integer data type could be used instead of a float. There are many designs that could be adopted in developing graphics capabilities of mobile devices both at the programming (algorithm) level and at the hardware architecture design level.

---

[3] These concepts will be more formally introduced in Chapter 3, but they are basically the memory portion containing the images to be rendered and the 3D depth information.

[4] So-called per-vertex operations convert the vertexes into geometric primitives.

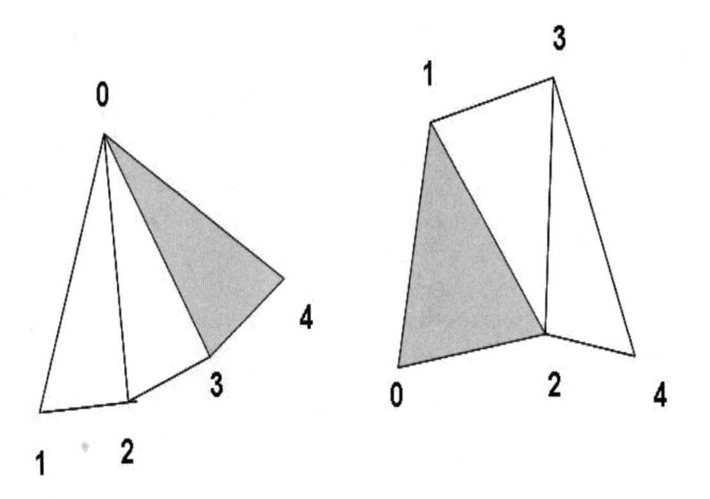

Fig. 2.7. Construction of Triangle Fan (left) and Triangle Strip (right).

2.3.1 Fixed-Point Maths

One the most useful and required features for a graphic application is to be able to represent and precisely manage nonintegral (real)[5] numbers. To represent rational numbers, it is necessary to codify and manage the decimal point position. There are mainly two solutions for this problem: the fixed point representation, where the number of decimal digits is fixed, and the floating point representation, where the number of decimal digits is not fixed. Instead, a pair of numbers are used called the *mantissa* and the *exponent*.

In early 3D graphics approaches, when computers had low computational capabilities, the fixed point representation was used, which was powerful enough for the graphics tasks even if not so precise in numerical approximation. This representation became obsolete when processors started to have increasing computational power and hardware units dedicated to floating point computations (also called FPUs), thus increasing the numerical precision.

Nowadays many high-level programming languages natively support the floating point representation. Mobile devices, instead, have all the issues described in this paragraph, and especially power consumption associated with graphics computations. And thus many of them don't natively support, or support only less performing, FPUs. For these reasons the fixed point representation is starting again to be widely used.

By considering N digits with fixed point representation it is assumed that the decimal point position is fixed in a location among the entire sequence. Thus we have a fixed number k of digits for the integer part of a number and $N - k$ for the rest of the fraction. To represent also negative numbers, the

[5] Natural rational numbers are ratios of integers.

most significative bit (MSB) of the integer part is used, as shown in Figure 2.8.

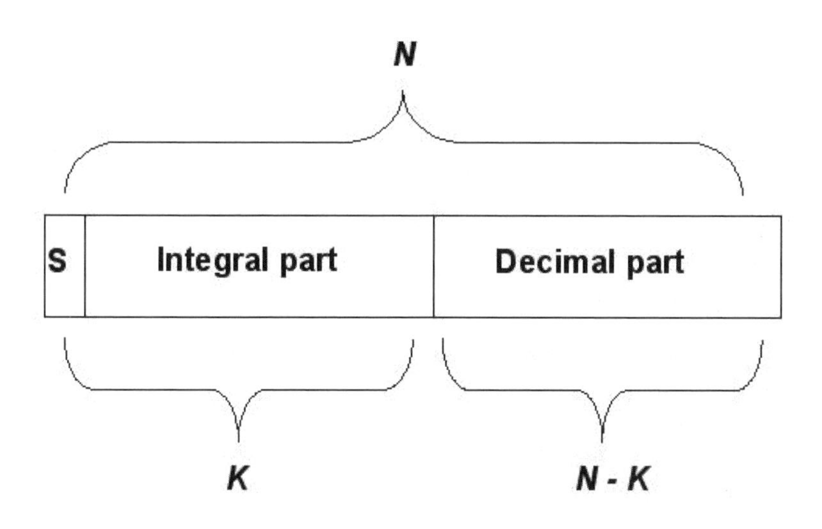

Fig. 2.8. Fixed point representation.

The set of different values representing numbers with the fixed point model, and having N digits, with p digits for the fraction, is $\{0, 2^{-p}, 2 * 2^{-p}, \ldots, (2^{N-1}) * 2^{-p}\}$.

It's clear that the granularity, that is, the interval between two consecutive numbers in this representation, is constant and is 2^{-p}.

It is a positional representation and thus the digits on the right of the decimal point are multiplied by negative exponentials of the base. We now describe two examples, based on powers of 2 and powers of 10, showing how the decimal part has weights with negative exponentials:

- $(5.75)_{10} = 5 * 10^0 + 7 * 10^{-1} + 5 * 10^{-2}$
- $(11.011)_2 = 1 * 2^1 + 1 * 2^0 + 0 * 2^{-1} + 1 * 2^{-2} + 1 * 2^{-3}$

Let's represent the number 22.315 in fixed point representation by using 8 bits, 3 of which are used for the fraction. The integer part could be converted by the successive divisions method; for the fractional part:

- The digits on the right of the decimal point are multiplied by 2.
- The integer part of the result is taken.
- Repeat the two preceding steps until the result is integer by itself.

Using this procedure the computations will be:

- $0.315 * 2 = \mathbf{0}.63$

- $0.63 * 2 = \mathbf{1}.26$
- $0.26 * 2 = \mathbf{0}.52$
- $0.52 * 2 = \mathbf{1}.04$
- $0.04 * 2 = \mathbf{0}.08$

Thus we will have $22_{10} \longrightarrow 10110_2$, for the integer part with successive divisions method; and $0.315_{10} = 0.01010\ldots_2$.

Since we could use only three digits as fixed point decimals in this example, the final result will be 10110.010.

Looking at the above example, it is clear how numbers have an approximated representation in the fixed point model that is strictly related to the number of digits used for the fractional part. At least in both the representation two kind of problems could emerge:

- *Overflow*: an error in the representation of a number (usually the result of operations) due to the fact the available number of digits is less than the ones needed to represent the number.
- *Undeflow*: the result is too small to be representable, thus is less than the smallest representable number.

2.3.2 Graphics Hardware Architecture Design

To illustrate the above-mentioned challenges we now describe two examples of architecture and programming solution to the above challenges.

A hardware architecture used for the triangles rasterization was proposed by T.A. Moller and J. Strom [29], in 2003. The rasterization process goal consists of identifying all the pixels stored in the frame buffer that belong to the geometric primitives. This solution provides a reasonable balance between visual rendering quality and requirements for system resources. In this new architecture there are three key innovations:

1. A new scheme for multisampling called FLIPQUAD Multisampling, which generates only two samples for each pixel and thus is better than the classical scheme [30].
2. A filter for textures, which implements a new compression algorithm for reducing texture size.
3. A simple *culling* [6] scheme, which avoids the need for rendering occluded surfaces.

Multisampling consists of a technique enabling *antialiasing* operations to be performed at full screen resolution. Antialiasing is a technique implemented

[6] In computer graphics, back-face culling determines whether a polygon of a graphical object is viewable to a viewing camera. It is a step in the graphical pipeline to test if the polygon is viewable or not. The process makes rendering objects quicker and more efficient by reducing the number of polygons for the program to draw.

to solve the problem of geometric shape approximation (low quality of images) caused by the discrete values assumed by pixel matrixes, as shown in Figure 2.9. These are, the square edged "jaggies" that exist on diagonal lines in relation to the square pixels that exist on the screen. The answer, termed antialiasing, is used to smudge those jaggies in order to create a smoother edge for objects. One process used to achieve antialiasing is called multisampling. The idea is that for each pixel, we sample the pixels around it to determine if this edge needs to be antialiased. Basically, we discard the jaggies by "smudging" the pixel itself.

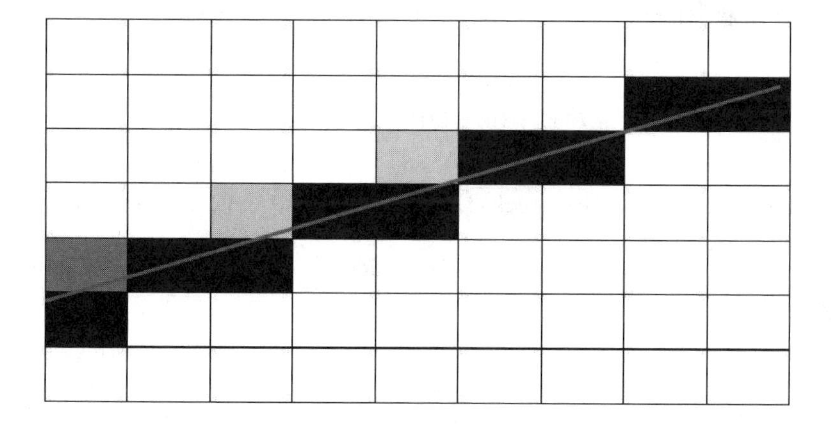

Fig. 2.9. Anti-aliasing approximations.

There basically exist two main approaches for full-screen antialiasing: supersampling and multisampling. Supersampling is based on a brute-force procedure. This approach samples an image to a higher resolution, usually doubled with respect to the original, and then computes the average of the 2×2 pixels (each pixel is doubled) and outputs it as the final result. The multisampling technique is less demanding in computational terms; in fact, every output pixel is made of a set of samples containing information on color depth. An average value is computed on the sample for each pixel in order to produce the output image. Both these techniques have to be implemented at the hardware level to provide acceptable performance and reduce the CPU work load.

The study produced by T.A. Moller and J. Strom [29] suggests a new multisampling scheme named *FLIPQUAD multisampling*, which uses two samples for each pixel and reorganizes them in patterns to obtain two different shading levels. The pattern used in this technique is shown in Figure 2.10.

The main advantages in using such pattern are:

- The four pixel samples are localized on x and y coordinates.

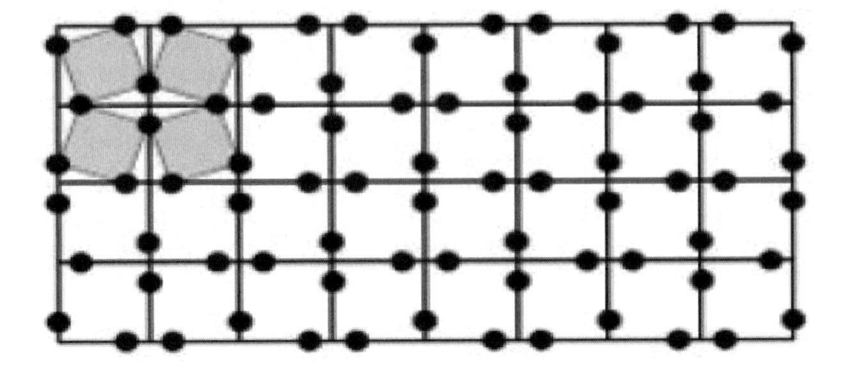

Fig. 2.10. FLIPQUAD pattern.

- The screen irregularity breaks the symmetry and thus increases the quality of the appearance.
- The sampling cost is only a little higher than that of the other existing schemes.

The texture compression algorithm has been enhanced, and this has become a starting point for a new texture compression system used on mobile devices, called PACKMAN [31]. The aim consists of a minimal compression scheme and a visual rendering quality of acceptable rate.

We will now briefly describe how this algorithm works: the input image is subdivided into 2×4 pixels blocks. Every block is represented by 32 bits. Only one color is supported for each block, and 12 bits of the 32 are used for storing red, green, and blue (R, G, B) components. Moreover, a study on human vision concluded that the human eye is more sensitive to the lighting than to the chromatic components of colors. The remaining 20 bits have thus been used for representing the pixel lighting for every block. For each pixel the base color is modified by a constant taken from a table of four elements. The same constant is added to the other color components. This approach requires two bits per pixel in order to specify which of the four values has been chosen. The remaining four bits are used as indices for the table specifying which table is used for the entire block. The decompression of each pixel takes place in the following manner:

- The block basic color is converted from 12 to 24 bits.
- By using the table index an entry in the table is chosen.
- The internal color is computed by modifying the 24 bits with the corresponding entry for the pixel in the table portion.

The hardware schema for the decompression is relatively simple. It is made of three 9-bit adders, a multiplexer for the modular components, and a table lookup capability.

Many hardware manufacturers have started to produce devices supporting these techniques using embedded chips.

2.3.3 Tile Rendering

A promising technique for data exchange with bandwidth reduction is called tile-based rendering (TBR). By reducing the data exchange, an important goal is achieved: power reduction. This technique approaches the problem of invisible surfaces by heavily modifying the sequence of operations in the rendering pipeline.

Figure 2.11 presents a classical rendering architecture.

The rendering process phases are as follows:

1. Geometric primitives are computed in any ordering and then a geometric transformation process takes place.
2. The transformed geometric data to be projected on the screen are then processed by clipping and rasterization parameter extraction phases.
3. In the rasterization phase a set of per-pixel computations are executed, such as z-buffer operations and color and lighting interpolations.
4. The invisible surfaces determination is delegated to the z-buffer.

Figure 2.12 presents an architecture based on tile rendering.

The tile rendering process steps are as follows:

1. Geometric primitives are computed in any ordering and then a geometric transformation process takes place.
2. The data to be projected on the screen are subdivided into tiles (portions) made of triangles, and for every computed portion the set of triangles intersecting it is added to a *triangle list*.
3. After processing all the geometries, we have a triangle list for every tile, and thus the rendering follows the execution in the list order, tile by tile.
4. For each tile the triangle list is matched against visible surfaces. Different methodologies could be used for pursuing this process, but in every case, since the tile is small (usually 32×32), it is possible to maintain in memory the portion of frame buffer and z-buffer for that tile.
5. After considering the triangle list only the visible triangles are rendered. Only the visible textures of triangles are loaded from memory.

Let's now compare the two processes, with attention to the bandwidth of data traffic. We use the term *data front* to indicate the data sent by the CPU or main memory to the graphic processor (geometric data, textures, and so on). With *data back*, we mean data transferred between the graphic processor and its own graphic memory. The study presented in [32] shows

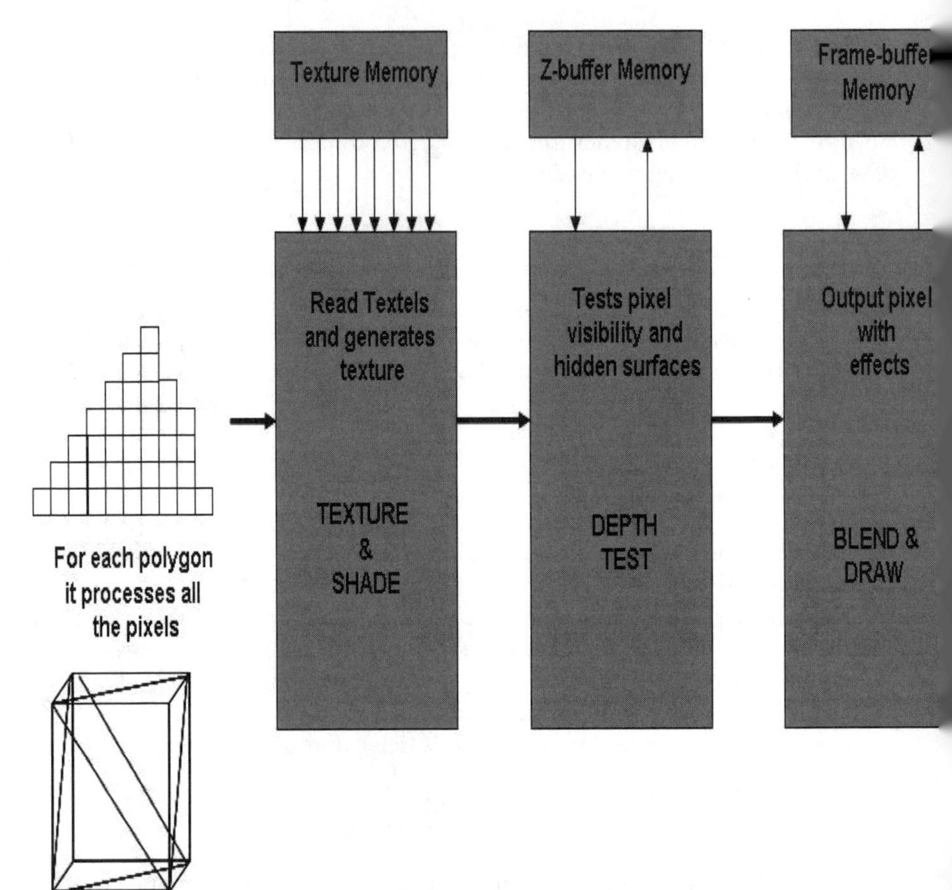

Fig. 2.11. Classical rendering pipeline.

within a classic architecture the amount of geometric data that is proportional to the number of used primitive, while in the tile-based architecture every geometric primitive could be sent to the *rasterizer* many times. For example, if a geometric primitive is present in n tiles, it should be transmitted n times to the rasterizer. Thus in a tile-rendering architecture, the data front component is strictly related to the number of tiles a primitive intersects, and this factor could be represented by the average number of tiles covered by a primitive.

While a tile renderer could increase the data front, the other component (data back) is significantly reduced. Since all data in a fragment[7], that belongs to a tile are memorized in a buffer (in the graphic processor memory), only the visible pixels (in the final image) should be written in the main memory buffer. In a traditional renderer many fragments are written anyway in the

[7] Every single pixel that has to be processed.

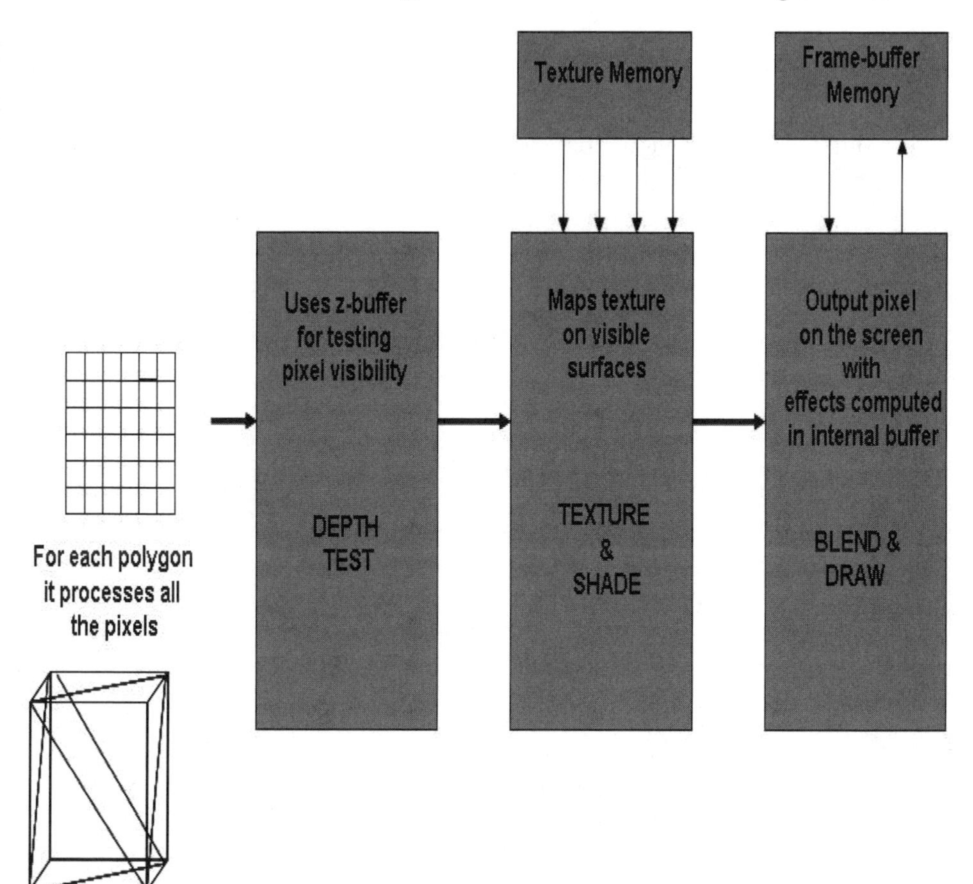

Fig. 2.12. Tile rendering architecture.

main memory buffer (which is more time-consuming to access), even if some of them could be occluded by others.

The experiment in [33] describes how the amount of external data traffic (main memory instead of graphic memory) changes with the tile size (in pixels). The result shows that 32 × 32 pixels seems to be a good size for tiles. If the tile size increases, the traffic is weakly reduced, while if the size is less than 32 × 32, the traffic strongly increases. Moreover, it has been measured how much data traffic to the external memory is reduced by using the tile rendering in comparison to the classic rendering. This second result shows how the new architecture reduces the external memory data traffic by 1.96 compared to the classic architecture.

Many other studies have been carried out on this technique, such as studies to efficiently determine when a primitive covers many tiles and which sorting algorithm works best with this technique [34].

This technique has been implemented in the PowerVR technology [PowerVR Mobile] by Imagination Technologies. It's actually a scalable architecture based on tile rendering, and it includes two different specifications for mobile devices, depending on whether they are set-top boxes, in-car devices, or PDAs and mobile phones.

2.4 Summary

This chapter presented what we believe are the more challenging and interesting scenarios in applications of three-dimensional mobile device graphics. We have identified three main areas of application, that many researchers, both in academia and in industry, are studying: mobile tourist guides, augmented reality, and mobile gaming. All these fields are challenging, and many steps are needed to evolve applications for better usability levels. The graphics technique and libraries we will describe in this book are, or will be, employed in the mentioned areas. Usability is a main issue for mobile devices, and with the increase in multimedia applications it becomes a very important topic. We explored usability among current devices, both describing the unresolved questions and the existent solutions. We then focused on new algorithms and architectural scenarios generated by new mobile device graphics capabilities. We presented recent results on architectural and algorithmic solutions for the three-dimensional scene rendering on mobile devices. These techniques are an initial step toward rethinking and optimizing programming code for the constraints of mobile devices. These architectures and algorithms highlight new problems in computer graphics that are stimulating the scientific and industrial community.

Part II

Mobile Graphics Programming

3

Introduction to Mobile 3D Graphics with OpenGL®ES

3.1 Introduction to OpenGL®ES

OpenGL®ES is a multiplatform set of API and it's also royalty-free; it includes a complete set of functions for 2D and 3D graphics on embedded systems like mobile devices, vehicles, and appliances. It is a subset of the well-known API set called OpenGL® (see Chapter 1), and it operates as an interface between the low-level software layer and graphics devices.

The main advantages of OpenGL ES are the following:

- **Standard and royalty-free.** Everyone can use the OpenGL ES specification and implement one's own version of these API. They are supported by many different hardware and software manufacturers because they are based on an open multiplatform standard.
- **Memory usage and power consumption.** "Embedded" devices (smart phones, palm tops, etc.) provide a different kind of hardware configuration from 400 MHz CPU and 64 MB of RAM typical of PDAs to 50 MHz CPU and 1 MB of RAM supported by mobile phones. OpenGL ES adapts by using only the memory required by each application, data/instructions, and throughput. In the very first implementation floating point operations weren't supported because of the lack of floating point units (FPUs) in embedded devices.
- **Extensible.** OpenGL ES includes an extensible mechanism that helps manufactures adapt it for newer hardware devices. Moreover, there is an OpenGL ES standard committee that evaluates and approves extensions to be included in the standard library.
- **User-friendly.** Based on standard specifications, the OpenGL ES library is already structured and developed with an intuitive design, and many hardware manufacturers are adapting to this new standard.

This chapter presents a set of examples to show 3D concepts readapted to the embedded environment, and provides source code examples of the API implementing these concepts. Appendix A shows how to setup the environment for supporting the OpenGL ES , which is very diffuse, but all the source code presented works with other implementations, because OpenGL ES is standard, except for different procedures for setting up the environments.

To understand the examples provided in this chapter, knowledge of basic C language is required; we won't focus on C but rather on OpenGL ES library, and thus a basic knowledge of the language will help. If the reader has had experience with OpenGL, it will be very simple to map this knowledge to the OpenGL ES API.

3.2 The OpenGL®ES Rendering Pipeline

We now focus on the OpenGL API structure, describing the steps needed to transform a geometric object into an image rendered on the screen. This process is called the *rendering pipeline.* The pipeline structure is used many times in this book to explain the basic concepts of rendering. All data are processed, transformed, and combined according to the OpenGL standard structure. The basic OpenGL instructions are still the same as the desktop OpenGL; only two elements have been changed:

1. Commands cannot be grouped in a *display list* to be further executed.
2. The first phase of the pipeline, used for approximating surfaces and geometric curves, has been discarded.

See Chapter 1 for further information on OpenGL and its pipeline.

3.3 3D Mobile Graphic Concepts and Rendering with OpenGL®ES

3.3.1 Starting with a Window

A basic library of functions is provided in OpenGL ES for specifying graphics primitives, attributes, geometric transformations, viewing transformations, and many other operations.

We will show generic code examples for the pipeline phases described above, and we will discuss how to set up a window for displaying the rendering results.

As we noted in the last section, OpenGL ES is designed to be hardware independent; therefore, many operations, such as input and output routines, are not included in the basic library. However, input and output routines and many additional functions are available in auxiliary libraries that have been developed for OpenGL programs.

The first step in developing an OpenGL application, and, in general, a graphic application, consists of setting up a display window.

For this purpose two basic libraries can be used: The first is called *libGLES_CM.lib*. The second is *ug.lib*, and is specific for abstracting the creation of a window and interface environment; this library abstracts from the OS implementation of the windows, thus behaving like the OpenGL utility toolkit (*GLUT*) for the standard OpenGL, and freeing us from supporting a specific OS. In fact, in addition to the OpenGL ES basic library, there are a number of associated libraries for handling special operations. The *ug.lib* provides a set of functions for interacting with any screen-windowing system. You'll see all the details for setting up a OpenGL ES environment in **Appendix A**.

It's possible to link these libraries using a specific integrated development environment (IDE), but we will use a more standard technique using the C*pragma* statement.

To link the library we can use the following syntax:

```
{#pragma comment(lib, "LIBRARY_NAME")}
```

In all graphics programs, we will need to include the header file for the OpenGL ES core library. For most applications we will also need the *ug.lib* for the visual interaction. For example, using the *ug.lib* for abstracting interfaces from the OS, we only need to include the file *GLES/gl.h*, as it includes the *GLES/egl.h* and all the needed OpenGL ES functions.

To create a graphic display context using OpenGL ES API, one must set up a *display window* on one's video screen. That is the rectangular area of the screen in which pictures will be displayed. We can't create the display window directly with the basic OpenGL functions, since this library contains only device-independent graphics functions, and window-management operations depend on the computer we are using A.1.

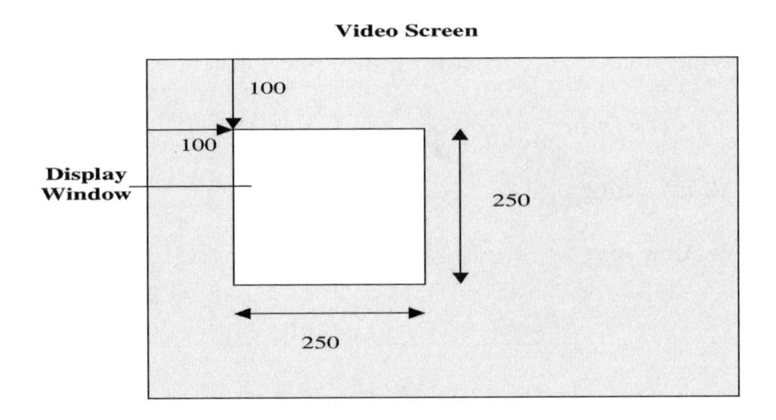

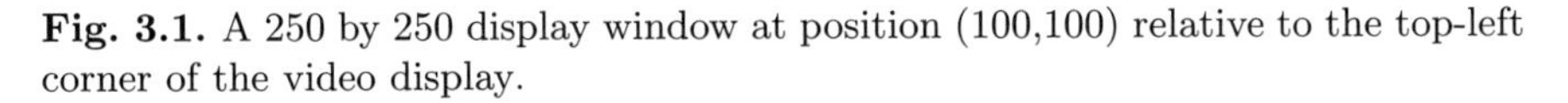

Fig. 3.1. A 250 by 250 display window at position (100,100) relative to the top-left corner of the video display.

The OpenGL paradigm displays graphic objects on the screen using frames. In fact, the performance is measured in frames per second (FPS). For each single frame it is necessary to develop what will be presented on the screen. The *display* function is executed for each frame; thus all graphics code will typically be inserted in this function. It requires an input parameter, which represents the displayed window.

The main routine should include an *init* function for initializing the graphic engine and returning an handle pointer for the graphic context.

The next step requires creating a window using a *create window* function. As already discussed for the *display* function, a window is used to store the handle of an OpenGL ES window. Typical function parameters are a graphic context handle, and a string with window title, height, width, and top left corner of the window.

At this stage we need to send to the OpenGL window the frame to be displayed. We can do this by using the *display* function, which takes as input a window and a display handle.

To prevent the program from immediately stopping, we need a *loop* function. This is what we actually call the *main loop.* This loop continues to iterate, managing the program messages and/or events. We can call a main loop function only with a parameter, a window handle.

The following generic steps create a simple window with OpenGL ES API. You'll find a detailed description in **Appendix A**.

```
int main() {

    GC gc = int(); // create a graphics context

    // create a window

    Window win = CreateWindow(gc, "Hello", 100,100,0,0);

    Display(win); // display function

    MainLoop(gc);

    return 0;

    }
```

Running this program does nothing, and that's why we haven't yet specified what geometric primitives to display on the window. We will now describe how to introduce basic interactions in OpenGL ES applications.

3.3.2 Basic Interaction

Many programs require inputting data from the keyboard or the mouse; thus we need a way to associate an action with a keyboard event.

The first step in managing keyboard inputs is to define a function that has some parameter input. We will define a basic one accepting four parameters:

1. The current window.
2. The key pressed, usually stored in an integer variable.
3. The x coordinate of the mouse pointer in the window when they key has been pressed.
4. The x coordinate of the mouse pointer in the window when they key has been pressed.

Then we will check which key has been pressed. This is usually implemented by a *switch* statement that associates the actions with the corresponding pressed key, for instance, "r", for the rotate action of fixed degrees, or "q" for quitting the application.

At this stage there should be a defined link from the keyboard input function to the main window in order to manage the keyboard events in that specific window.

3.3.3 Geometric Primitives and Per-Vertex Operations

OpenGL ES includes a set of basic geometric primitives for representing geometric shapes in rendering process: points, lines, triangles, and so on. Every primitive is characterized by a type and the number of vertices that can be minimized by using certain specific primitives like (*strip* and *fan*),as shown in Figure 3.2.

For a triangle strip, for instance, one must specify three vertices for the first triangle; then only one vertex for each subsequent triangle is required. For each vertex, it is necessary to compute the position with respect to viewpoints, using geometric transformations, and considering the characteristics of displayed window. Note that only a matrix (as you'll see in the next examples) expresses the coordinate transformations from the model space to the point of view of the user; this matrix is usually called the *model-view* matrix. Initially the *model-view* matrix is an identity matrix, but then it is continuously multiplied by the transformation matrices (rotations, scaling, ...). Thus every transformation like rotations or scaling is related to the state of the model-view matrix.

We will present some code fragments for different models of views and basic geometric primitives that are rendered as per-vertices operations, in order to show how graphics objects are mapped in the rendering pipeline.

There are basically two graphic projections supported by OpenGL ES : orthographic and perspective. If you ever looked down a very long road, you probably noticed that the road seems to be smaller (in width) depending on

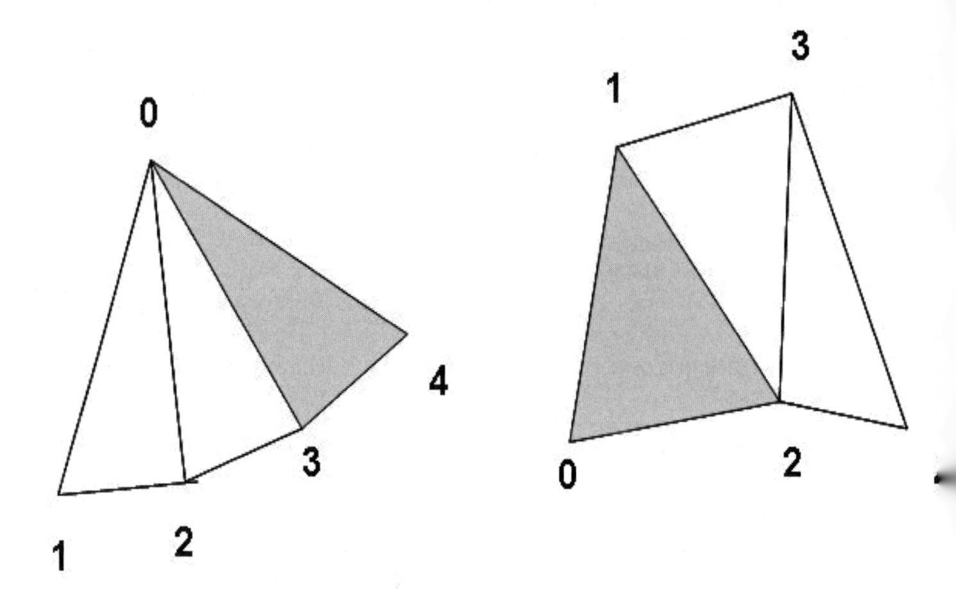

Fig. 3.2. An example of triangle strips.

the distance from you (the user point of view). This is called a perspective view. In the orthographic view, the road remains the same size independent of the distance of the viewpoints. We will describe how to display a shape on the screen using an orthographic view. This is only an example, to show you how geometric primitives work and to show the per-vertex operations done in the rendering pipeline (we will develop example code abstracted from the hardware that implements the rendering pipeline).

The orthographic view is less computationally expensive than is perspective, and thus could be preferred to the perspective view, especially on mobile devices. In fact often there is no need for perspective, as in 2D applications or video games. In all those cases the orthographic view is very useful (Figure 3.3).

Geometric primitive shapes, like squares or triangles, are implemented by specifying the vertices of their geometric shape. Vertices are points in a three-dimensional space, and thus are made of three coordinates: x, y, and z. After specifying the vertices, one must specify the type of geometric primitive as shown in the following table:

OpenGL ES provides the **glPointSize** to change points size and the **glLineWidth** for lines size. it is usually assumed that a line segment is one pixel thick.

For example, if we decide to display a square shape, we need to specify coordinates for vertices of the square. Usually this is done by three float values representing the x, y, and z coordinates of a vertex. The x stands for

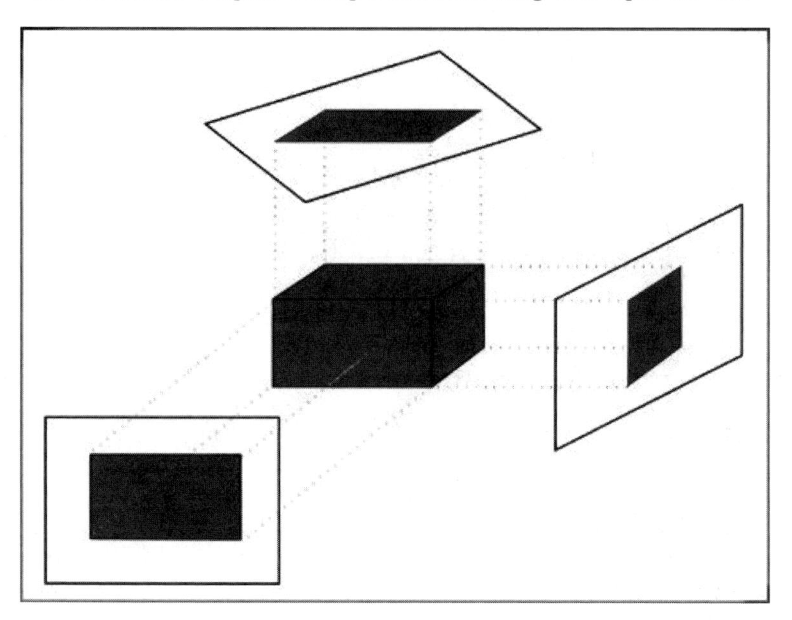

Fig. 3.3. Orthographic parallel projections.

Geometric primitive type	Output
GL_POINTS	A point for each vertex
GL_LINES	A line for every couple of vertices
GL_LINE_STRIP	Given the first vertex, a line joining all the vertices in the given sequence is displayed
GL_LINE_LOOP	The same as GL_LINE_STRIP, but the last vertex is conjoined to the first one
GL_TRIANGLES	For every three vertices a triangle is displayed
GL_TRIANGLE_STRIP	After the first two vertices, every successive vertex uses the previous two vertices to draw a triangle
GL_TRIANGLE_FAN	After the first two vertices, every successive vertex uses the preceding and the first vertex to draw a triangle; it is used for conic shapes

Table 3.1. Geometric primitives.

the horizontal position, the y is vertical, and z stands for depth. The greater the z value, the more the vertex will be shown close to the user viewpoint, while if z is negative, the vertex will be displayed far from the viewpoint. The square will be represented by showing two adjacent triangles. The first three vertices could be used to draw the first triangle, while the last is used for the final triangle. We can store all data into an array, as shown in the following box:

```
GLfloat square[] = {


    x_{1t1}, y_{1t1}, z_{1t1},
    x_{2t1}, y_{2t1}, z_{2t1},
    x_{3t1}, y_{3t1}, z_{3t1},
    x_{1t2}, y_{1t2}, z_{1t3},


    };
```

($x_{it1}, y_{it1}, z_{it1}$ is a vertex) of the first triangle and ($x_{1t2}, y_{1t2}, z_{1t2}$) is a vertex of the second triangle.

The graphic window is initialized by declaring its background with the function **glClearColor**. This function takes four parameters as input, with values between 0 and 1. These parameters identify a single color in red, green, blue, and alpha transparency (RGBA) color space. The first three values represent, respectively, red, green, and, blue, while the fourth stands for the alpha channel, i.e., the transparency.

As already described, we usually need to introduce an initialization function; following a sketch code of the *init* function, it is presented and commented on.

```
void init{}{
    glClearColor(r,g,b,a);
    glMatrixMode(GL_PROJECTION);
    glLoadIdentity();
    glOrthof(left, right, bottom, top, near, far);
    glVertexPointer(3, GL_FLOAT, stride, square);
    glEnableClientState(GL_VERTEX_ARRAY);
    }
```

Many kinds of matrices can be used working with the OpenGL ES API. They usually define the projections or geometric primitives to be used; transformations are managed by the *GL_MODELVIEW* matrix, whereas projections are managed by the *GL_PROJECTION*. The current matrix can be changed by using the function **glMatrixMode**; which takes as input a projection matrix.

We then initialize this matrix with the identity matrix, the values of which are set to 1; we use it to clear the matrix.

To compute the orthogonal projection views, we use the **glOrthof** function. It requires six parameters (Figure 3.4):

- GLfloat **left** and GLfloat **right**: values for clipping (intersecting) planes on the left and right.
- Glfloat **bottom** and GLfloat **top**: values for clipping (intersecting) planes on top and bottom.
- GLfloat **near** and GLfloat **far**: values for planes containing the scene; geometric primitives closer than *near* or more distant than *far* won't be displayed.

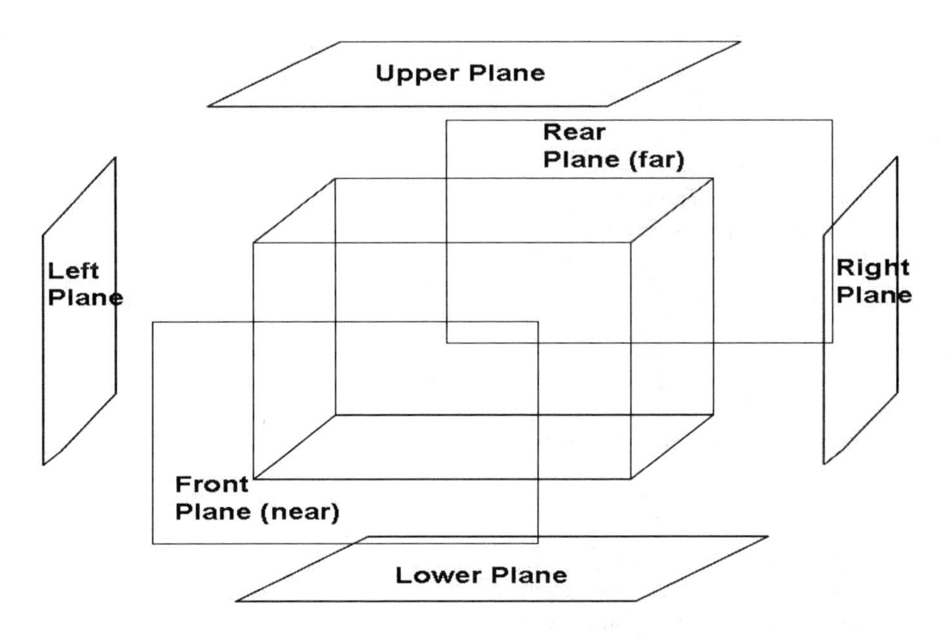

Fig. 3.4. A visual example of the six parameters.

The parameters described above span the area to be visualized on the window.

After setting up the orthographic view matrix we consider the functions needed to draw the square. Geometric primitives in OpenGL ES are displayed by *Vertex Arrays*. To use those arrays, we select vertices, and this is done by the **glVertexPointer** function. This function takes as input four parameters:

- GLint **size**: specifies the number of coordinates per vertex. Since each vertex has been defined by using three values, this parameter is usually set to 3.
- GLenum **type**: represents the array data, e.g., GL_BYTE (bytes), GL_SHORT (integers), GL_FLOAT (floating), etc.
- GLsizei **stride**: specifies the offset between consecutive vertices, i.e., how many values there are between the end of one vertex and the start of the next. In our example this value is set to 0.

- const GLvoid ***pointer**: defines a memory address of the first element of the array, i.e., the array pointer.

Like the vertex array, there are many other specialized arrays included in the OpenGL ES library; if they aren't used by our code, we can disable them thus recovering resources for the rest of the code. By default all these arrays are disabled and thus if we need to use them, we have to enable them. This can be done by the **glEnableClientState** function, which takes arrays as the input parameter. In our example it is the GL_VERTEX_ARRAY.

After setting up the selected projection (orthographic) and the geometric shape (a square made of two triangles), we can display it.

The following sketch of code shows a typical display function:

```
void display(win) {
     ClearScreen(win);
     glDrawArrays(mode, first, count);
     glFlush();
     SwapBuffers(win);
     }
```

To draw the geometric primitives using the current array, we use the **glDrawArrays** function, which takes the following parameters as input:

- GLEnum **mode**: defines which type of primitive to display, in our case GL_TRIANGLE_STRIP.
- GLint **first**: specifies the first index of the array, and thus how many vertices skip before starting to read array values, since we want to start from the first it will be set to 0.
- GLsizei **count**: defines how many vertices to read; in our case four (remember the square is defined by two triangles specified by a triangle strip with four vertices: three for the first triangle and one for the second).

We send data to the screen by flushing and swapping. The final displayed image will look like Figure 3.5.

We will now describe another example related to per-vertex operations: how to apply transformations to geometric primitives.

OpenGL ES provides three basic primitive transformations:

1. Scaling
2. Translation
3. Rotation

Scaling allows the geometric primitives to increase or decrease in size without changing their proportions.

Translation moves a geometric primitive along the three axes of the 3D space.

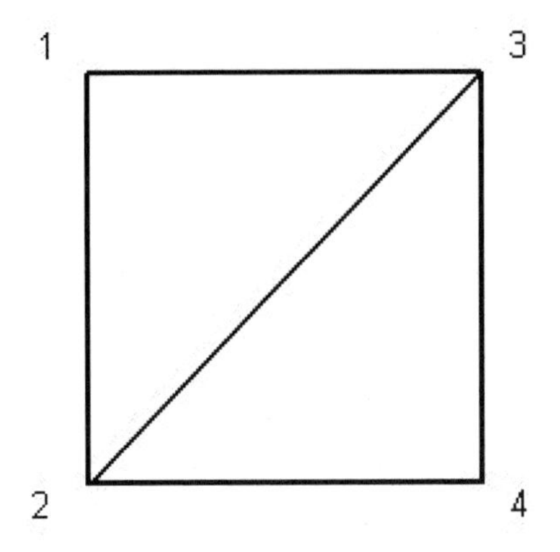

Fig. 3.5. An orthogonal view of a square made by two triangles stripped together.

Rotation turns a geometric primitive with respect to the three axes.

To show a rotation (as an example of transformation) we will sketch the code of a rotated triangle. We will then introduce an array containing a triangle vertex as already seen in the previous example.

```
GLfloat triangle[] = {

    x_{1}, y_{1}, z_{1},
    x_{2}, y_{2}, z_{2},
    x_{3}, y_{3}, z_{3},


    };
```

We now define a color array. Colors are specified by groups of four coordinates in RGBA color space as in the example above.

```
GLfloat colors[] = {

    r_{1}, g_{1}, b_{1}, a_{1}
    r_{2}, g_{2}, b_{2}, a_{2}
    r_{3}, g_{3}, b_{3}, a_{3}


    };
```

In the initialization function we simply set the background color.

```
void init() {

    glClearColor(r,g,b,a);

}
```

The display function code will look like this:

```
void display() {
    ClearScreen(color);
    glShadeModel(GL_SMOOTH);
    glVertexPointer(3, GL_FLOAT, stride, triangle);
    glColorPointer(4, GL_FLOAT, stride, colors);
    glEnableClientState(GL_VERTEX_ARRAY,
                        GL_COLOR_ARRAY);
    glPushMatrix();
    glTranslate(x, y, 0);
    glScale(0.5, 0.5, 0.5);
    glRotate( rot.degree, x, y, z);
    glDrawArrays(GL_TRIANGLES, first, count);
    glPopMatrix();
    glFlush();
    SwapBuffers(win);
    }
```

The code draws a triangle including a smooth color. There are two kinds of shading models implemented by the **glShadeModel** function: GL_FLAT and GL_SMOOTH. By default it is set to GL_SMOOTH. GL_FLAT sets a single color for the shape to be represented (Figure 3.6). GL_SMOOTH enables the so-called smooth shading; that is, vertices and primitive colors are computed by interpolating single values of the vertex' colors (Figure 3.7).

Transformations are executed by **glTranslatef**, **glScalef**, and **glRotatef**. The *f* at the end of each function name indicates that the functions accept only floating point (real) parameters as input. Usually an x indicates a GLfixed type for input parameters, while v is used for arrays (vectors).

After drawing the triangle, we don't want other shapes to be influenced by the next transformations, since that will change geometric coordinates. The **glPushMatrix** and **glPopMatrix** functions are used for storing the current status of the coordinate reference system in a stack. We will insert our transformation code between these two functions so that the *model-view* matrix will be reset to its original status (reference system) after the code segment.

Fig. 3.6. The triangle with flat shading.

Fig. 3.7. The triangle with smooth shading.

We then apply all three transformations. Each transformation will change the *model-view* matrix, thus influencing subsequent transformations. Recalling that geometric transformations are carried out by matrix multiplication, the product in matrix algebra changes with the order of the factors. In our example, we translate, rotate, and scale the shape.

glTranslate takes three parameters as input, thus indicating how to move along the three axes. Our first transformation consists of moving the triangle only by x, y coordinates.

A scaling function takes three values for each vertex and multiplies them for its input parameters. If the shape was placed in the bottom left of the window, the operation will have scaled the object but the origin will remain centered in the bottom left corner.

We reduce our triangular shape by half with the *glScale* function.

Finally, rotation takes place. The first parameter specifies a rotation angle. The other three are used for indicating with respect to which axes the object is rotated.

We draw the triangle with *glDrawArrays.* Then, we restore the original reference system by taking out from the stack the original matrix, using the *glPopMatrix.*

3.3.4 Lighting

In the OpenGL ES lighting model for each primitive vertex, the corresponding color is computed considering the attributes of the selected material and lights. The vertex lighting information requires defining the geometric normals in order to manage object reflections for the selected vertex. OpenGL ES separates lighting effects in three RGB coordinates; every lighting source is characterized by a quantity of red, green, and blue (expressed within 0 and 1) light it emits, and every material is defined by the percentage of reflected color. Moreover, for each color there are four lighting types, as described in Chapter 1 [11].

There are three light colors for each light - ambient, diffuse, and specular (set with glLight) - and four for each surface (set with glMaterial). All OpenGL implementations embed at least eight light sources, and glMaterial can be changed for each polygon.

The definitive polygon color is the sum of all four light components, each of which is shaped by multiplying glMaterial color by glLight color (changed by the directionality in the case of diffuse and specular). Whereas there is no emission, color for glLight is added to the final color without changing it.

For example, the diffuse lighting component changes the color of a vertex in RGB by computing:

```
[max(L*n,0)]*(MR_{diff}*LR_{diff}, MG_{diff}*LG_{diff},
MB_{diff}*LB_{diff}),
```

where $L * n$ is the scalar product between the unitary vector L from the vertex to the lighting source and the unitary normal n of the vertex. All these parameters are combined together in an equation for resolving the color to be stored in the frame buffer for future *per-fragment* operations.

The OpenGL ES lighting model is an estimate of lighting physics, and it doesn't, for instance, include second-order reflections or shadows mixing (the shadow of an object crosses the shadow of another), but this is not our goal, since the OpenGL ES goal is to be capable of executing *real-time* rendering even with limited resources.

Before starting to provide a code example for managing lighting, let's recall the geometric normal concept given in Mathematics, which be valuable for the OpenGL ES lighting model [35].

To display light we must compute normal vectors for each polygon in an object.

A normal of a polygon is a perpendicular vector connecting to the polygons surface, and it is very useful for frequent implementations of 3D computer graphics when considering surface direction mechanics.

Since all models in a 3D scene will be made out of polygons, it is convenient to have a function that calculates the normal vector of a polygon. A normal vector of a polygon is the dot product of two vectors located on the surface plane of the polygon (in our case that polygon is a triangle). And what we need to do is take any two vectors located on the polygon's plane and calculate their dot product. The dot product is the resulting normal vector. The OpenGL ES library provides a **glNormal3f** function for defining normals.

We will sketch the code for supporting two kinds of lighting, by defining two color arrays - one for ambient light and another for diffuse light.

```
float lightambient[] = { r_{a}, b_{a}, g_{a}, a_{a} };

float lightdiffuse[] = { r_{d}, b_{d}, g_{d}, a_{d} };
```

An array for specifying material properties is also needed - one for ambient and another for the diffuse light. Basically we multiply lighting values by material values to obtain a final reflected color. Each value represents a quantity used for reflecting a particular color.

```
float materialambient[] = { r_{ma}, b_{ma}, g_{ma},
                             a_{ma} };

float materialdiffuse[] = { r_{md}, b_{md}, g_{md},
                             a_{md} };
```

We sketch the code for the *init* function as follows:

```
void init() {

glEnable(GL_LIGHTING);

glEnable(GL_LIGHTx);

glMaterial(GL_FRONT, GL_AMBIENT, materialambient);

glMaterial(GL_FRONT, GL_DIFFUSE, materialdiffuse);

glLight(GL_LIGHTx, GL_AMBIENT, lightambient);

glLight(GL_LIGHTx, GL_DIFFUSE, lightdiffuse);

glClearColor(r, g, b, a);

glVertexPointer(3, GL_FLOAT, stride, triangle);

glEnableClientState(GL_VERTEX_ARRAY);

glShadeModel(GL_SMOOTH);

}
```

We activate lighting by using the GL_LIGHTING parameter as input to the **glEnable** function.

OpenGL ES allows the use of eight different lights at the same time. To enable one of these lights, a GL_LIGHTx parameter has to be passed to the **glEnable** function as input, with $x = 0 \ldots 7$.

To define material properties we use **glMaterialfv** and **glMaterialf** functions. **glMaterialfv** is used for multiple valued parameters, while the **glMaterialf** is used when there is a single parameter, as shown later in this example.

The first parameter defines which polygon face needs to be updated by lighting information. In OpenGL ES API, GL_FRONT_AND_BACK flag is available. The second parameter is used to specify the type of lighting attributes and it can be GL_AMBIENT, GL_DIFFUSE, GL_SPECULAR, GL_EMISSION, or GL_AMBIENT_AND_DIFFUSE.

The last parameter is an array or single value depending on the selected function (**glMaterialfv** or **glMaterialf**).

Lighting properties have to be set, and this is done by using **glLightfv** and **glLightf** functions, which work in the same manner as the material functions.

A sketch code for the display function includes the computation of normals for managing lights.

```
void display(win) {
    ClearScreen(color);
    glLoadIdentity();

    //FRONT AND BACK
    glColor(r, g, b, a);
    glNormal(0.0f, 0.0f, 1.0f);
    glDrawArrays(GL_TRIANGLE_STRIP, first_{front},
    count_{front});
    glNormal(0.0f, 0.0f, -1.0f);
    glDrawArrays(GL_TRIANGLE_STRIP, first_{back},
    count_{back});

    //LEFT AND RIGHT
    glColor(r, g, b, a);
    glNormal(-1.0f, 0.0f, 0.0f);
    glDrawArrays}(GL_TRIANGLE_STRIP, first_{left},
    count_{left});
    glNormal}(1.0f, 0.0f, 0.0f);
    glDrawArrays(GL_TRIANGLE_STRIP, first_{right},
    count_{right});

    //TOP AND BOTTOM
    glColor(r, g, b, a);
    glNormal(0.0f, 1.0f, 0.0f);
    glDrawArrays}(GL_TRIANGLE_STRIP, first_{top},
    count_{top});
    glNormal(0.0f, -1.0f, 0.0f);
    glDrawArrays(GL_TRIANGL_STRIP, first_{bottom},
    count_{bottom});
    glFlush();
    SwapBuffers(win);
}
```

Normals must be perpendicular to surfaces. Thus the surface in front of the light must have a *(0,0,1)* normal vector, while the back surface has *(0,0,-1)*. The vector length is one; thus both are normalized vectors.

Normals are defined by the **glNormal3F** function applied before drawing the related primitive, and this function takes as input three parameters that identify a normalized vector.

The same thing is done for the bottom and side surfaces. Like the color and vertex arrays, there is also a normal array. It could be initialized as a **glNormalPointer** function, which works like the **glVertexPointer**.

To enable this array, a GL_NORMAL_ARRAY flag must be passed to **glEnableClientState**.

Figure 3.8 shows the visual feedback of lighting with color features.

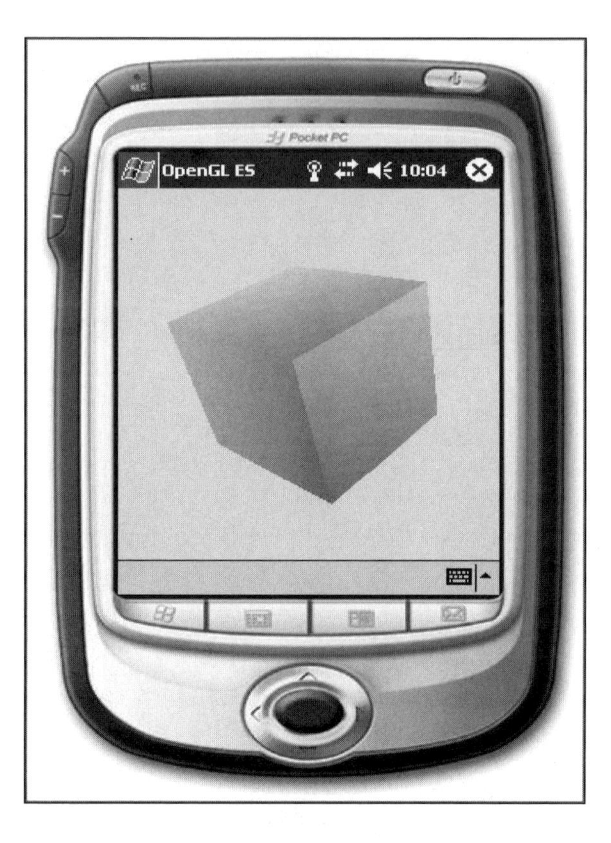

Fig. 3.8. A scene with the color shading enabled.

In the first sketch of code we enrich our scene by including lighting. But the lights didn't have a particular direction. We will see now how to use directional lights; this will allow us to manage diffuse and specular illumination.

First let's create arrays for setting light properties, and add a specular array for managing specular effect.

```
float lightambient[] = { r_{a}, b_{a}, g_{a}, a_{a} };
float lightdiffuse[] = { r_{d}, b_{d}, g_{d}, a_{d} };
float lightspecular[] = { r_{s}, b_{s}, g_{s}, a_{s} };
```

We then create a specular array for materials. We set it so that the material will reflect all the light that hits it.

```
float materialambient[] = { 1.0f, 1.0f, 1.0f, 1.0f };

float materialdiffuse[] = { 1.0f, 1.0f, 1.0f, 1.0f };

float materialspecular[] = { 1.0f, 1.0f, 1.0f, 1.0f };
```

Since we selected a directional light, we must set light position and direction. We create two arrays for specifying these two properties.

```
float lightposition[] = { x_{p}, y_{p}, z_{p} };

float lightdirection[] = { x_{d}, y_{d}, z_{d} };
```

We now sketch the *init* function:

```
void init() {
    glEnable(GL_LIGHTING);

    glEnable(GL_LIGHTx);

    glMaterial(GL_FRONT_AND_BACK, GL_AMBIENT,
    materialambient);

    glMaterial}(GL_FRONT_AND_BACK, GL_DIFFUSE,
    materialdiffuse);

    glMaterial}(GL_FRONT_AND_BACK, GL_SPECULAR,
    materialspecular);

    glMaterial}(GL_FRONT_AND_BACK, GL_SHININESS,
    shininess);

    glLight(GL_LIGHT0, GL_AMBIENT,
    lightambient);

    glLight(GL_LIGHT0, GL_DIFFUSE,
    lightdiffuse);
```

```
glLight(GL_LIGHT0, GL_SPECULAR,
lightspecular);

glLightfv(GL_LIGHT0, GL_POSITION, l
ightposition);

glLightfv(GL_LIGHT0, GL_SPOT_DIRECTION,
lightdirection);

glLightf(GL_LIGHTx, GL_SPOT_CUTOFF, angle);

glLightf}(GL_LIGHTx, GL_SPOT_EXPONENT, exponent);

glClearColor(r, g, b, a);

glShadeModel(GL_SMOOTH);
}
```

We enable lighting and the first light; then we set the material properties and specular values.

We then set a new material property by using the **glMaterial** function. The shininess value for the material is usually in the $[0, 128]$ range. This value specifies how much a specular light will be polarized. The greater the value, the more the light will be polarized.

The next step consists of setting the light properties. To set the position and direction of the lights; the *GL_POSITION* and *GL_SPOT_DIRECTION* flags must be set and passed as input to **glLightfv** function.

Another useful flag is *GL_SPOT_CUTOFF*. It specifies a light cone size. We can imagine an effect that is like an electric torch cone pointing to a wall. For instance, a value of 180 will spread light in every direction.

The *GL_SPOT_EXPONENT* is used to specify how polarized a light will be. For example, we could think about a torch that concentrates its light when turned toward a direction. As for *GL_SHININESS* flag, this value can be in the range of 0 to 128.

Finally, there are three more flags that could be used:

- *GL_CONSTANT_ATTENUATION*
- *GL_LINEAR_ATTENUATION*
- *GL_QUADRATIC_ATTENUATION*

They can be used to manage light reduction which is the measure of how much the light intensity is reduced by moving far from the light source. Thinking of the torch example, the effect is light reduction when moving far from the torch itself. Setting these properties could end in decreasing software performance since they require many computations and thus we won't use them in our example.

We create a sphere by using the **SolidSphere** function. We use h horizontal and v vertical slices for lighting effects.

The sketch of code for the display function follows:

```
void display(win) {

    ClearScreen(color);

    glLoadIdentity();

    SolidSphere(size, h, v);

    glFlush();

    SwapBuffers(win);
    }
```

The final scene rendering will display a sphere with a specular reflection (Figure 3.9).

Fig. 3.9. A sphere with a specular reflection.

3.3.5 Per-Pixel Operations and Texture Mapping

Data can be input in the form of pixels rather than vertices. In fact, for pixel operations, pixel data are either:

- Stored as texture memory, for use in the rasterization stage.
- Rasterized, with resulting fragments merged into a frame buffer as if they were generated from geometric data.

Color information contained in pixels is converted into appropriate format to be assigned to processed vertices. For example, if pixels are taken from an image stored at 24 bits of color depth and the display is capable of visualizing only 2^{16} colors, then an 8-bit-per-channel transformation takes place with the $5-6-5$ bits scheme, respectively, for red, green, and blue components (16 bits all in RGB5 format).

Strings of bits, called *texels*, are then stored in texture memory and processed frame by frame. As we will see in next examples, textures need to be sampled at least one time per frame in order to be usable; this process has a relevant impact on data bus bandwidth. One possible solution to this problem consists of compressing textures in memory and then decompressing them when required. Using this process less memory, disk space and bandwidth are required, obtaining a fast rendering procedure; otherwise more textures could be stored in the same memory space, or higher resolution textures could be saved in memory.

Texels are specified by two coordinates in a bidimensional texture image (which is created with height and width lengths that are exponentials of base 2). By mapping these coordinates in 3D space, it is possible to detect which texels have to be used for coloring geometric primitive pixels on the screen. Moreover, textures are managed according to the right perspective of geometric primitives. They need to be matched against the object's dimension in pixels; a matching operation is called *magnification* if texels are less than the destination dimension in pixels, and it is called *minification* otherwise.

We describe now how to introduce texture mapping on geometric polygons. The first phase consists of loading textures from an external file. Texture supported file formats are mainly bmp, jpg, gif, and png, but also other formats are supported. We will focus on the *bmp* file format, or bit map, since it's quite simple to write code for loading a bit-map formatted image.

Once a texture has been loaded from memory, it doesn't matter in what format it was originally stored. It will be saved in memory. Note that OpenGL ES works with images of sizes like 64×64, 128×128, and so on; they're all in the form of $2^n * 2^n$.

This section covers the basics of texture mapping in OpenGL ES API. This includes uploading textures in memory and the application of texture onto geometry. We do not cover the actual loading of the texture data itself. That will be shown in detail in **Appendix A**.

We have some raw RGB image data in an array and we want to load it for our geometry in OpenGL ES API. The first thing we have to do before OpenGL ES can use this raw texture data is upload it to the video memory. Once a texture is to uploaded to the video memory, it can be used throughout the time in which our application is running. Before a texture can be uploaded to the video memory, there is some setup that must take place. Below we show a sketch for functions calls needed for uploading and drawing textures. Note that these function calls should be used once per texture [36].

The first thing to do in the process of uploading textures is calling the **glBindTexture** function. **glBindTexture** specifies which texture "ID" will be used to point at the texture. A texture "ID" is defined as a number that could be used in order to access our textures. Following we show a snippet of code for **glBindTexture** function.

```
void init (){
     glEnable(GL_TEXTURE_2D); }

void loadtexture () {
     glBindTexture(...,  id);
     glPixelStorei (...);
     glTexParameteri}(...);
     glTexEnvf}(...);
     glTexImage2D(GL_TEXTURE_2D, 0, GL_RGB, width, height,
     0, GL_RGB, GL_UNSIGNED_BYTE, data);
     }

void drawtexture () {
     glBindTexture}(...,  id);
     glBegin (...);
     glVertex}(...);
     glTexCoord}(...);
     glEnd (...);

}
```

This call will set textures that have the identifier *id* as the active texture. Any other call that has to manage texture mapping will affect this texture.

The **glPixelStorei** function specifies how data need to be uploaded. For instance, if the parameter is 1, the function specifies that the pixel data are aligned in byte order, that is, the data have one byte for each component - one for red, green, and blue.

glTexParameteri sets parameters for the current texture.

The **glTexEnvf** function specifies environment variables for the current texture. It denotes how texture will behave when it is rendered into a scene. It sets active textures by a *modulate* attribute. The *modulate* attribute enables us to apply effects such as lighting and coloring to textures.

The **glTexImage2D** function will upload textures to video memory, making them available for use in our programs. We will describe parameters for this function:

- *target*: the target of this function, GL_TEXTURE_2D in our case.
- *level*: the detail level as a number, which could be 0 for this snippet.
- *internal format*: internal components parameter. This parameter indicates how many color components to memorize from the uploaded texture. There are symbolic constant values for this parameter, but generally GL_RGB is used.
- *width* and *height*: the width and height of a bit-map image. These must be integers equal to $2n + 2$ for some integer n. The textures width and height must be a power of two.
- *border*: image borders, which must be 0 or 1. The value 0 specifies not to use image borders.
- *format*: the pixel format that will be uploaded. There are many constants that can be used, but again GL_RGB is the value that is mostly used.
- *type*: type of data to be uploaded. Usually GL_UNSIGNED_BYTE is used.
- *pixels*: pointer to image data. This is the image structure that will be loaded to video memory.

We uploaded our texture and we want to use these data in video memory, for example to draw textures on top of three-dimensional shapes on the screen. The process for applying a texture to geometric shapes depends on data types that we have to manage. We thus have to deal with texture coordinates and types, as can be seen in the *draw-texture* function snippet of code (Figure 3.10).

First we need to be sure that texturing is enabled. We can do this by the *glEnable* (GL_TEXTURE_2D) function. We then specify a texture coordinate for each vertex that is part of a face. As shown in the *draw-texture* function, a pattern for texture mapping is like the following: *TexCoord*, and *VertexCoord* (Figure 3.11).

3.3.6 Per-Fragment Operations

Per-fragment operations are mainly devoted to enhancing the graphic appearance of 3D applications, but they also are involved in object depth tests. In fact, by using the *z-buffer* algorithm, these operations can determine if a surface is visible or not (occluded by others). One typical example of these operations is the *blending* phase, which facilitates representing transparencies by mixing objects and background colors, by means of a special kind of equation involving many variable parameters.

Another important effect managed by per-fragment operations is *antialiasing*, which smooths object contours. When a scene may give rise to *moir* patterns (when the original image is finely textured) or jagged outlines (when the original has sharp contrasting edges, e.g., screen fonts), antialiasing techniques are used to reduce such artifacts [37] [38].

The *fogging* effect is also managed at this operational level of the pipeline and makes objects that are further from the camera progressively more obscured by haze. This technique works because of light scattering, which causes

Fig. 3.10. The image above shows the OpenGL texture coordinate system. In the code above, the calls to glTexCoord2f are very important as to what the end result of the texture mapping will be. When we make a call to glTexCoord2f (x,y), OpenGL places texture coordinates at that place on the image. If we are texturing a triangle, there will be three texture coordinates on the image. Once a glEnd is reached, the triangle that is formed by the texture coordinates is then mapped onto the triangle that is made up from the vertices.

more distant objects to appear hazier to the eye, especially in outdoor environments. The fragment colors are interpolated by the following equation:

- $C = f \times C_p + (1 - f \times C_f)$

where C is the computed color, C_p is the starting color, f is a density coefficient, and C_f if the fog color. If f is linearly dependent on the distance, the perception of distance effect is simulated.

We now treat a simple case of blending with a snippet of code to show an example of a per-fragment operation. Blending occurs after the scene has been rasterized and converted to fragment, but before the final computed pixels are drawn in the frame buffer. With blending, we can control how much object color values can be combined with new fragment values for using alpha blending to create a translucent fragment.

Enable blending is made by *glEnable*(GL_BLEND) function, while setting up different blending modes involves *glBlendFunc*(source, destination) function, with two parameters as the source and destination.

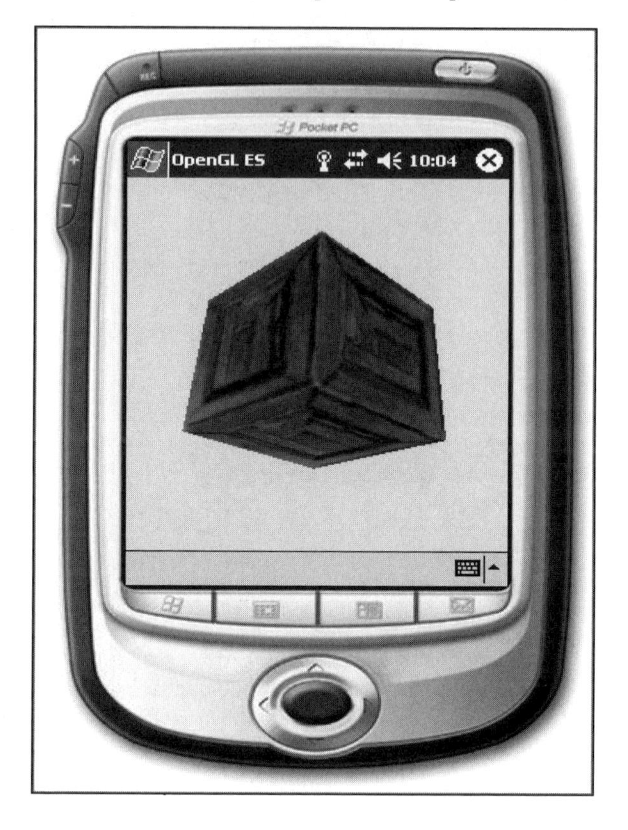

Fig. 3.11. Texture mapping.

The tables below show the parameter values that could be set with blending functions.

Constant	Blend Factors
ZERO	$(0, 0, 0, 0)$
ONE	$(1, 1, 1, 1)$
DST_COLOR	R_d, G_d, B_d, A_d
ONE_MINUS_DST_COLOR	$(1, 1, 1, 1) - (R_d, G_d, B_d, A_d)$
SRC_ALPHA	A_s, A_s, A_s, A_s
ONE_MINUS_SRC_ALPHA	$(1, 1, 1, 1) - (A_s, A_s, A_s, A_s)$
DST_ALPHA	A_d, A_d, A_d, A_d
ONE_MINUS_DST_ALPHA	$(1, 1, 1, 1) - (A_d, A_d, A_d, A_d)$
SRC_ALPHA_SATURATE	$(f, f, f, 1)$, $f = min(A_s, 1 - A_d)$

Table 3.2. Values controlling the source blending function and the source blending values they compute [39].

Constant	Blend Factors
ZERO	$(0,0,0,0)$
ONE	$(1,1,1,1)$
SRC_COLOR	R_s, G_s, B_s, A_s
ONE_MINUS_SRC_COLOR	$(1,1,1,1) - (R_s, G_s, B_s, A_s)$
SRC_ALPHA	A_s, A_s, A_s, A_s
ONE_MINUS_SRC_ALPHA	$(1,1,1,1) - (A_s, A_s, A_s, A_s)$
DST_ALPHA	A_d, A_d, A_d, A_d
ONE_MINUS_DST_ALPHA	$(1,1,1,1) - (A_d, A_d, A_d, A_d)$

Table 3.3. Values controlling the destination blending function and the destination blending values they compute [39]

We consider blending operations by taking RGB components of a fragment as representing its color, and the alpha component as representing transparencies. For example, if we are viewing an object through red glasses, the color we see is a component of red from the glasses and a component of the object color. The percentage of mixing varies according to the transmission properties of the glasses. If glasses transmits 65 percent of the light (that is, 35 percent opaque), the color we see is a combination of 35 percent of the glass color and 65 percent of the color of objects we are looking at. We consider that situations including multiple translucent surfaces can also happen. If we look at one street from a driver's point of view and the car has the windshield between it and the viewpoint, and we are wearing sunglasses, objects behind the car are visible through two pieces of glass.

A typical example is in the snippet of code sketched below.

```
int DrawScene() {
    ClearScreen(color);
    glLoadIdentity(); // Reset the Current
                      // Model-view Matrix
    glEnable(GL_BLEND); // Enable Blending
    glBindTexture(..., id1);
    glBegin(...);
    glVertex(...);
    glTexCoord(...);
    glEnd(...);

    glEnable(GL_BLEND); // Enable Blending
    glBindTexture(..., id2);
    glBegin(...);
    glVertex(...);
```

```
glTexCoord(...);
glEnd(...);
// we need at least two textures to blend
// Set the Blending to 50 percent modality
glBlendFunc(GL_ONE, GL_ONE);
glDisable(GL_BLEND); // Disable Blending

return 0;
}
```

3.4 OpenGL®ES Future Developments and Extensions.

Recent developments in graphics hardware have replaced fixed modules of rendering pipeline with programmable modules. The so-called *shading language* has already been introduced in standard OpenGL, and it allow developers to program and change some phases of the graphics pipeline. The developer provides single, independently compiling software units called *shaders*, while there is a main program that links together the individual modules. There are two languages involving *vertex* or *fragment* depending on the corresponding pipeline phase. A vertex processor is a pipeline module that is in charge of vertex values and geometric operations, as already seen. Operations executed by this module (which at the hardware level usually include also the per-pixel operations) are:

- Vertex transformations
- Normals computation and transformations
- Texture coordinates generation
- Texture coordinates transformations
- Lighting
- Color on materials

Software units written for this module are called the *vertex shader*.

The fragment processor is a programmable module that operates on fragments and corresponding values. Operations supported by this module are:

- Interpolation and blending
- Texture loading and applying
- Fogging
- Linear operations on colors

Software units written for this module are called the *fragment shader*.

The very first version of OpenGL ES 1.0 has been based on OpenGL 1.3, but removing some data types and adding smaller data types and fixed point math.

We now describe some of the OpenGL ES extensions. If the reader is already familiar with OpenGL can see the changes; otherwise it is a short list of differences to start with after learning the standard OpenGL:

- **OES_byte_coordinates**: allows and supports byte arrays for vertices and textures.
- **OES_fixed_point**: for each function having a float as a parameter there is another accepting a fixed point input and its name ends with an x, for instance, *glClearColorx*.
- **OES_single_precision**: since the double data type is not supported, this extension replaces every function having double type parameters with new ones, for example, *glFrustum*.
- **OES_read_format**: enables developers to specify the input format for the *glReadPixels* operation. Before, the only possible input format was GL_RGBA or GL_UNSIGNED_BYTE.
- **OES_compressed_paletted_texture**: supports the use of textures made of 16 or 255 color palettes in five different formats (24-bit RGB, 32-bit RGBA, RGB565, RGB5551, RGB4444), and can be used by the *glCompressedTextImage2D* command.
- **OES_query_matrix**: allows applications to read the current values of matrices in the stack.

Moreover, new extensions are coming out for enhancing visual rendering and performances, like the support for the *vertex buffer object* (VBO). *Vertex buffer objects* support is crucial for rendering objects based on vertex arrays. The idea behind VBOs is to provide regions of memory (buffers) accessible through identifiers. A buffer is made active through binding, following the same pattern as other OpenGL entities such as display lists or textures.

VBOs provide control over mappings and unmappings of buffer objects and define the usage type of the buffers. This allows graphics drivers to optimize internal memory management and choose the best type of memory, such as cached/uncached system memory or graphics memory in which to store the buffers.

Binding operations convert each pointer in client-state functions into offsets relative to current bound buffers. As a result, bind operation turns a client-state function into a server-state function. The scope of data used by client-state functions is only accessible by OpenGL client, and other OpenGL clients are not able to access that client data. Because the VBO mechanism changes client-state functions into server-state functions, it is now possible to share VBOs data among various clients. As a result, OpenGL clients are able to bind common buffers in the same way as textures or display lists.

A sketch of code for supporting VBOs is shown below.

```
//some data
GLfloat data[] = {

    x_{1}, y_{1}, z_{1},
    x_{2}, y_{2}, z_{2},
    x_{3}, y_{3}, z_{3},


    };

GLuint buffer_obj;
//the classic gen and bind used from OpenGL
// creates and selects the buffer object
glGenBuffers(1, &buffer_obj);

glBindBuffer(GL_ARRAY_BUFFER, buffer_obj);
//store data into buffer object, which is in
//graphic memory when available

glBufferData(GL_ARRAY_BUFFER, sizeof(data),
data, GL_STATIC_DRAW);

//select vertex buffer object
glBindBuffer(G_ARRAY_BUFFER, buffer_obj);
//use it like a normal vertex array and draw as triangle
glVertexPointer(3, GL_FLOAT, 0, 0);

glDrawArrays}(GL_TRIANGLES, 0, 3);
```

Finally, OpenGL ES also includes a specification of a common platform interface layer, called *EGL*. This layer is platform independent and may optionally be included as part of a vendor's OpenGL ES distribution. The platform binding also has an associated conformance test. Alternatively, a vendor may choose to define its own platform-specific embedding layer.

3.5 Summary

In this chapter we explored the OpenGL ES library, describing the rendering pipeline associated with it. We then introduced many of computer graphics' basic concepts, such as: windows and mouse interaction and geometric primitives. We described all the concepts involved in the rendering pipeline by giving examples with OpenGL ES API. We provided snippets of code for geometric primitives and per-vertex operations, lighting managing, per-pixel operations and texture mapping, and finally per-fragment operations. We defined concepts and then provided sketch code for their implementation within the OpenGL ES library; for complete working code examples, refer to Appendix A. We concluded our OpenGL ES description by describing future developments, like *vertex shaders*, which are a powerful extension to OpenGL ES API. We provided also a pseudo-code for the shaders in order to show users their abilities to change the rendering pipeline, thus revealing a powerful tool for optimization.

4

Java™Mobile 3D Graphics

4.1 M3G

This section introduces the Mobile 3D Graphics API, M3G also known as JSR-184[1] [40].

Even if a 3D Graphics API for Java already exists (JAVA3D), most mobile devices have limited memory and processor power; thus Java 3D is unsuitable for them. Therefore, a proposal for a more suitable API was put together by a group of experts. The need was for a scalable, small-footprint, interactive 3D API for mobile devices that could work as an optional package for J2ME™to allow 3D graphics. Java Platform, Micro Edition or Java ME (formerly referred to as Java 2 Platform, Micro Edition or J2ME), is a collection of Java API for the development of software for resource constrained-devices such as PDAs, cell phones, and other consumer appliances. M3G is a software package for providing 3D graphic functionalities to a wide range of devices(Figure 4.1).

M3G is designed to be a 3D API suitable for the J2ME platform and CLDC [41]/MIDP [42].

Since it uses floats, it cannot be implemented on top of CLDC 1.0 but must be implemented on at least version 1.1 of CLDC. The Connected Limited Device Configuration (CLDC) is a specification of a framework for Java ME applications targeted at devices with very limited resources such as pagers and mobile phones. It could possibly be implemented on MIDP[2] 1.0, but most devices supporting M3G will likely also support MIDP 2.0 [37]. It is integrated with components of MIDP to allow efficient rendering to its *Image* and *Canvas* classes.

---

[1] Java™Specification Requests (JSRs) are formal documents that describe proposed specifications and technologies to be added to the Java platform.

[2] Mobile Information Device Profile (MIDP) is a specification published for the use of Java on embedded devices.

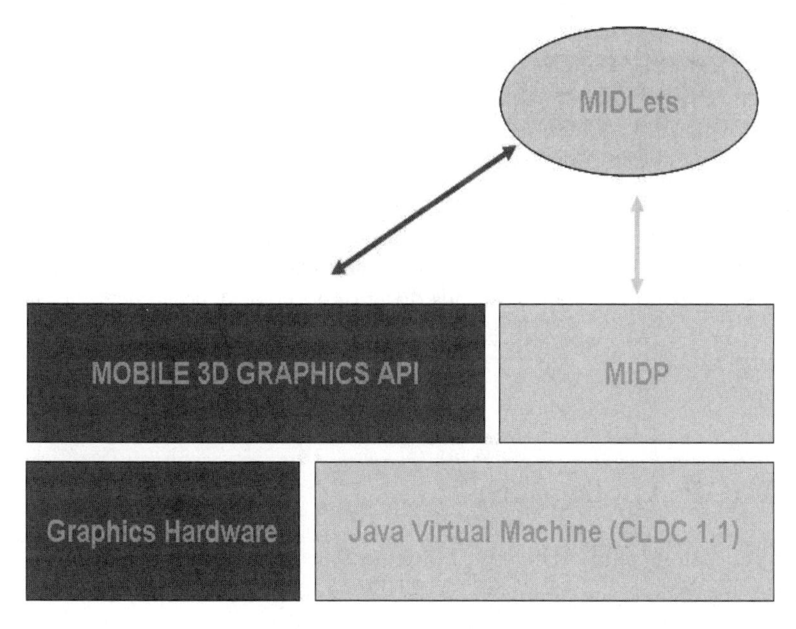

Fig. 4.1. Mobile 3D Graphics architecture.

We consider, now, the question about a need for a new mobile 3D Graphics standard when OpenGL ES is already available.

OpenGL ES is a low-level API standard, since it's based upon OpenGL; in fact, even for building simple 3D scenes it requires developing many lines of code, while there's a need for having a compact version of the final application. M3G is a high-level library designed to be compatible with OpenGL ES API. It is not a competitor, but it's more a complement to the OpenGL ES API set. This design choice has many advantages:

- Enhances developer's productivity.
- Minimizes code size of graphics applications.
- Increases applications' performance.

JSR-184 must not be mixed up with Java 3D standard API, which extends 3D capabilities to standard Java applications. Java 3D was designed for desktop computers and its completely unsuited for M3G. Many solutions and technicalities used in Java 3D has been modified and reused in JSR-184, as for example the support for a *scene graph*, which is used for representing in a compact and hierarchical structure all the elements that are part of a 3D scene. The *scene graph* represents a tree structure and includes definitions of each kind of physical or abstract object in the 3D world (cameras, lights, animations, etc.). The root of this scene tree is in fact represented by a *World node* object.

Moreover, JSR-184 specifications describe a new standard file format (*.m3g*), used for including all data related to a specific scene (the scene graph) and loading these data in applications coded to support the M3G standard. In this way, the scene data, including animations, can be created by using common 3D modeling programs (Maya, 3D Studio, ...) available on the market. These models can then be saved in M3G format and imported in a M3G application program that, by using few lines of code, can visualize and animate the imported scene. The product life cycle is thus tremendously accelerated by a clear separation between graphics design and code development of applications. In fact, graphic artists can create their own look and feel for the scene, including animations, and then export them as an M3G file to application developers.

4.2 MIDP Applications

The applications supporting the M3G standard are MIDP applications, and they are also known as *MIDlets*. The MIDP specification defines the minimum hardware, software, and network requirements for an application to run on a certain kind of device. MIDP applications are co-resident with other applications and executed in the *Mobile Information Device* (MID) framework, as shown in Figure 4.2.

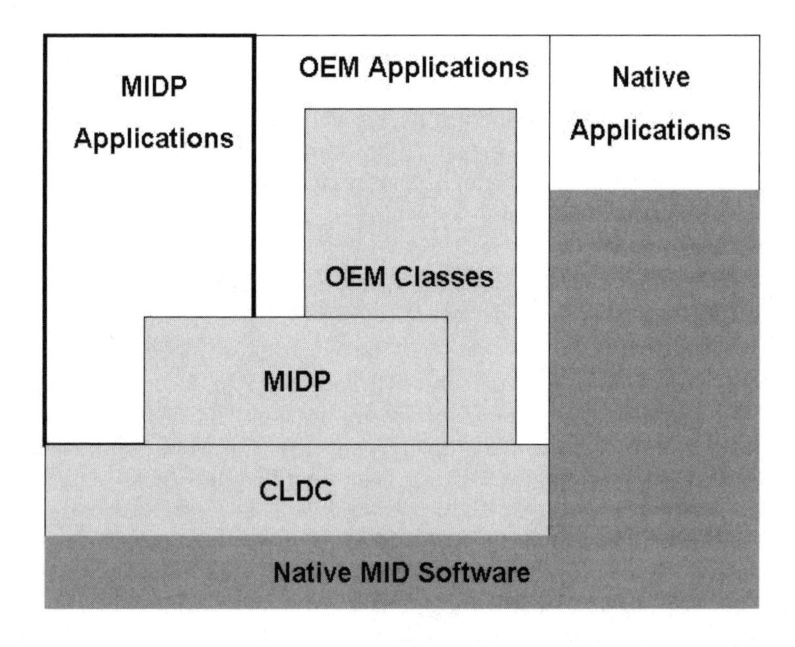

Fig. 4.2. MIDP architecture.

We describe now some basic concepts useful for designing and implementing a MIDP application. A life cycle of a MIDlet application is shown in Figure 4.3.

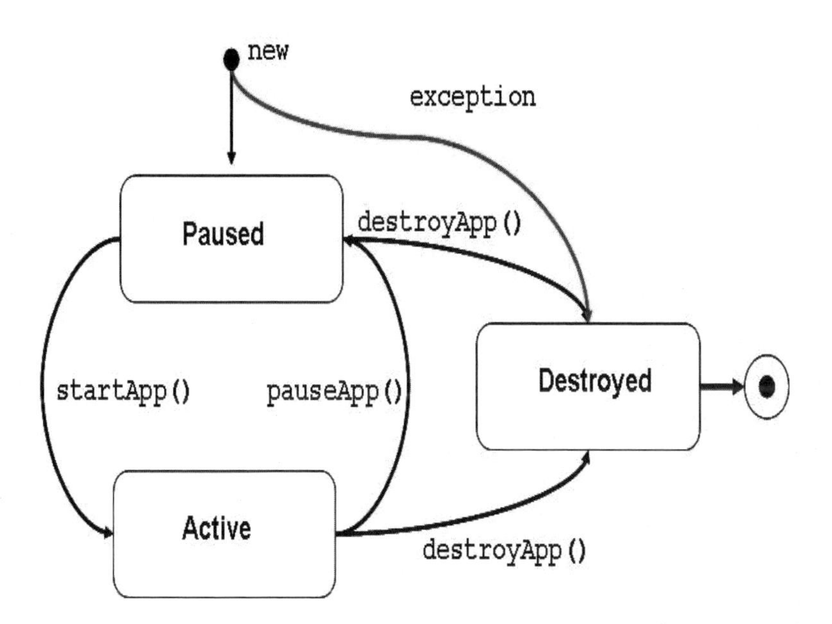

Fig. 4.3. MIDlet life cycle.

Application Management Software (AMS) is an environment where a MIDlet is installed, executed, stopped, and uninstalled. AMS creates every new instance of a MIDlet and manages its status during the execution process. A generic MIDlet can be in one of the following statuses: *Paused*, *Active*, *Destroyed*. When created and initialized, it is in the *Paused* status; if there is an exception raised by MIDlet constructor, it goes into *Destroyed* status. MIDlet goes into the *Active* status when a call of the *startAPP()* method is completed.

Figure 4.3 shows the statuses of a MIDlet and the functions that manage transitions from one status to another.

The user interface in MIDP applications is built by using two API functions. The first API, which is a low-level function, is extended from the abstract class *Canvas*, and the second, which is a high-level function, uses *Alert*, *Form*, *List*, and *TextBox* classes, extended from a *Screen* abstract class as shown in Figure 4.4.

High-level API classes are designed to provide portability among software components on different MIDs. The *Canvas* class allows applications to have direct control over the user interface but delegates to programmers porta-

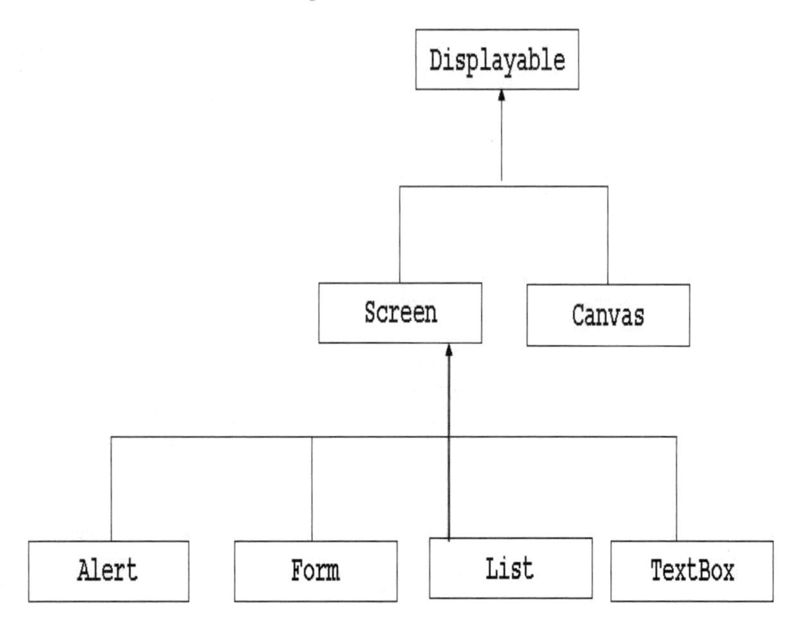

Fig. 4.4. High-level MIDP classes.

bility implementation among MIDs interfaces (display size, supported colors number, different kinds of keyboards, etc.).

Usually, one or more MIDlets are packed in a *JAR* file called a *MIDlet suite*, which is then used by AMS. MIDlets included in the same suite share the same execution environment (virtual machine) and thus can interact among one another. Every MIDlet suite can be associated with an application descriptor used for describing its content. The descriptor file extension should be *.jad*. It is used for managing MIDlets and storing application configuration properties. These properties can be modified in a *JAD* file without changing the associated *JAR*. MIDP specification provides detailed information on building and developing application descriptors and their attributes.

In the latest versions of MIDP standard, there are some interesting new features:

- HTTP secure protocol support (HTTPS)
- Enhanced network management
- Support for application distribution
- New graphical user interface (GUI) components
- Gaming support
- Multimedia functions support (audio and animations)
- A security model (trusted MIDlets)

The main enhancement is a package entirely dedicated to gaming development for J2ME framework, called *javax.microedition.lcdui.game*. It includes several classes that enable game developing for mobile devices [43]. In particular, a *GameCanvas* class can be used in conjunction with an M3G standard on devices supporting MIDP version 2.0.

4.3 Immediate and Retained Mode

The main class for drawing a scene with M3G standard is the *Graphics3D* class. It is defined as a singleton [3] and a unique instance can be accessed via the *getInstance()* method. To draw a scene, it is necessary to link a *Graphics3D* instance to a *target* object, draw the scene by an appropriate method, and release the *target*, as shown in the following snippet of code.

```
Graphics 3D g3g = Graphics3d.getInstance();
World = world; ...
Graphics g = ...
boolean bound = false;
try {
    g3d.bindTarget(g);
    bound = true;
    g3d.render(world);
}
finally {
    if(bound) g3d.releaseTarget();
}
```

A target object is a common *Graphics* object, the same as used in the *paint()* method with a *Canvas* or a *GameCanvas* class.

Graphics3D can also draw on top of an Image2D object. In this way a developer can draw a three-dimensional scene and use it as texture. Note that target objects must be released after using them; otherwise a unique instance of Graphics3D cannot be linked to other objects, and buffers cannot be sent to the screen for visualization.

Graphics3D supports two different drawing modalities:

1. *Immediate mode*
   - This is a low-level modality that allows defining each detail of a drawing process.
   - It draws an individual node, a group of nodes, or a submesh in a scene graph.
   - Cameras, lights, and background are managed separately.
2. *Retained mode*

[3] A one-time instantiated class with a unique point of access.

- This mode hides low-level details by loading and visualizing three-dimensional scenes by means of a few lines of code.
- It directly draws the *World* object, at the root of a scene graph.
- It manages cameras, lights, and background by accessing them directly with a *World* object.

The *retained mode* allows developers to use already-made, complex, three dimensional models; for instance, a developer can easily manage a *scene graph* in order to build a car model. Nodes representing wheels can rotate around their axes and are constrained to be parallel with respect to the car body orientation. All this information can be used by specifying it during modeling as additional information to nodes. The *retained mode* simplifies 3D world design by hiding low-level technical details from developers.

The overall control of a 3D scene can be obtained only by using low-level functions, and by accessing the graphics pipeline, and thus, for this reason, JSR-184 supports also the *immediate mode*, where drawing functions could be invoked on single objects. Moreover, the *retained mode* can take advantage of graphics acceleration because it is built on low-level *immediate mode* functions. Both modalities can be used in conjunction with each other, allowing developers to balance drawing performance with resources by choosing the appropriate modality with respect to their target.

4.4 Scene Graph

The *retained mode* uses a *scene graph* for linking all geometric objects in a three-dimensional world made of a tree structure. Each node of the graph represents a geometric object and contains information on appearance, [4] positioning in space, and function with respect to other nodes.

To build a 3D world, objects are used as subclasses of the *Node* base class. Then the *Group* class contains many objects, and the *World* class is a special case of the group class that includes all nodes in a scene. A *World* node is root of the scene graph and it is different from a regular node, in that all specified transformations are ignored during scene rendering.

A 3D world can be created from scratch, and new nodes can be linked after that, but a more convenient procedure is storing a scene in an *.m3g* file and then loading that scene to manage it in a scene graph. A complete and basic *scene graph* includes at least a *World* object and a *Camera* object. Figure 4.5 shows a generic *scene graph* describing different node characteristics.

It also possible to share components among different nodes of a *scene graph*, thus reducing memory usage. There are, anyway, some exceptions:

- Nodes can belong only to one group.
- Cycles among nodes are not allowed.

[4] It includes all geometric information concerning the node.

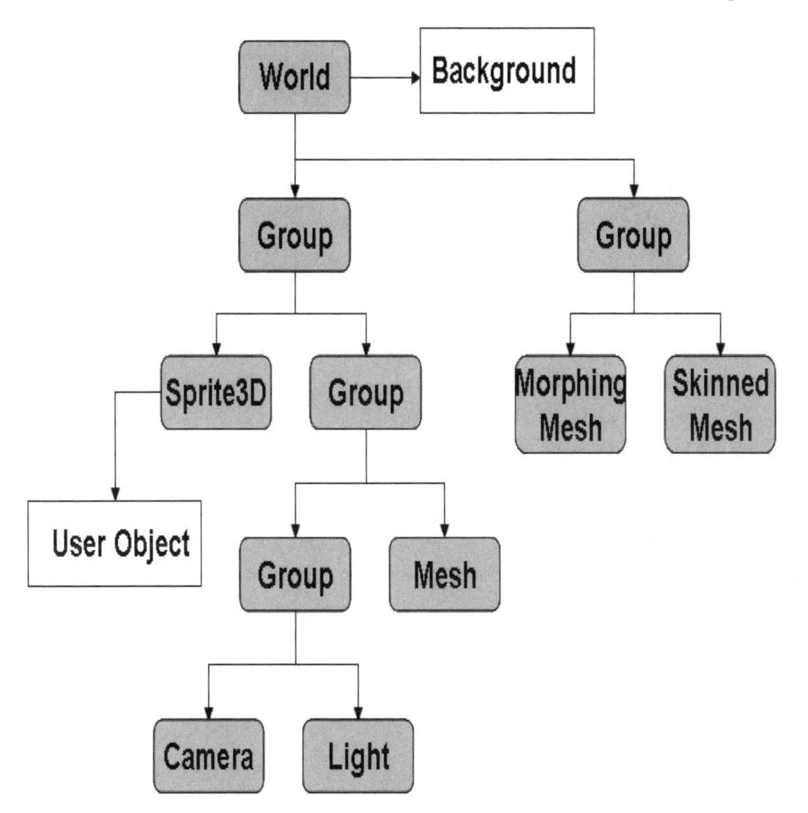

Fig. 4.5. A scene graph example.

To uniquely identify and address an object within a scene graph, a field of *Object3D* class is used, the *userID*. Each node in a tree holds its own *userID*, while *World* object has 0 by default; all other objects can have arbitrary values. It is important to observe that *userID* values are not unique within a tree, and so different nodes could have the same *userID*.

To build a scene graph from an *.m3g* file, the *Loader* class is invoked; it manages object extraction from files and builds all necessary classes. It creates all animation controllers, and it initializes the tree structure and all group of nodes, lights, and cameras. All these functionalities are included in a single *load()* method. More specifically, all classes stored in the *.m3g* file are deserialized in a vector containing *Object3D* objects returned as the result of a method call.

For example, in order to use a first element of an *M3G* file called *test.m3g*, as an instance of a World class, we need to write the following code:

```
Object3D[] o = null;

try {
    o = Loader.load("test.m3g");
}

catch (Exception e) {}

World loaderWorld = (World) o[0];
```

This class can also load many types of image formats, such as *PNG*, and in this case the result of the load method would be an *Image2D* object.

4.5 Transformations

The abstract class *Transformable* defines all geometric transformations that can be applied on nodes. There are four types of transformations:

- Translations (T)
- Rotations (R)
- Nonuniform scale (S).
- Generic nonhomogeneous matrix 4×4 (M)

Given a point in the space $p = (x, y, z, w)$, representing a vertex coordinate or a texture coordinate, its transformation can be defined (with respect to a coordinate system) as follows:

$p' = TRSM \times p$

A *Transformable* class defines methods for setting these components, also individually, as for example with the methods *setTranslation()* or *setScale()*.

4.6 Nodes of the Scene Graph

The *node* class is an abstract class representing all kinds of nodes included in a *scene graph*, such as: lights, cameras, meshes, sprites, and groups. A node defines a local coordinate system that can be transformed with respect to its ancestor coordinates system. Nodes can also be lined up with other nodes or point to a reference node; in this way we can force, for instance, a light or a camera node to point to a fixed object.

Another interesting characteristic of a node is the *ScopeID* parameter. This field is used for setting the visibility levels of a node, and in general is used for computing the visibility of a set of objects. Many different kinds of masks can be defined for the visibility of parts of a scene and to modify the scope of the camera in order to match the parts of a scene that are visible. If the scope of a camera and the nodes do not match, the nodes aren't drawn on the screen, thus saving resources especially at the computational level.

Moreover, this parameter can be used for speeding up computations on lighting. Usually, in a three-dimensional environment, the lights have a certain radius, determined by the type of light and its intensity. By setting different scopes for lights and objects corresponding to their distances, it can be computed if a light has effects on that object or not. This allows the use of many different light sources in the same scene without affecting the speed of performances and saving computational resources.

A set of nodes can be grouped together by using a *Group* class. Grouping different objects can help in the case of managing different objects with the same kind of operations. A typical group example is a car model with four wheels. In fact, by defining a car as a group of nodes, it is possible to move the whole car without moving each wheel individually.

4.7 Camera Class

A camera class is represented by a node in a scene graph, which sets the position of observers in the scene and the projection of a 3D perspective on a two-dimensional display.

The camera is generally pointing toward negative values of the z axis. It can be positioned and oriented in the same way as other nodes, namely by using transformations available at each node. It uses classical projections and clipping rules that apply for *OpenGL*, with the exception of the user-defined clipping planes, which are not supported. It is, instead, possible to define many cameras, and thus it is posssible to have many different viewpoints.

4.8 Managing Illumination

The *JSR-184* specification supports four kinds of lights, each having different computational complexity and thus performance. The equations used for light computation are directly imported from the *OpenGL* standard ones. Light types are:

- *Ambient light*: defines the general intensity of objects in a scene. Ambient lights illuminate a scene with the same illumination quantity; thus position and direction are ignored during computations.
- *Directional light*: defines only the source direction of light. Position or distance from an object has no effect on the latter, even if it can be set anywhere in the scene.

- *Omni light*: defines a light source point. The omni lights affect objects in each direction. A curve can be set to adjust the intensity variable according to the distance from objects.
- *Spot light*: defines the position, direction, and radius of a light cone. This light doesn't have any effect on objects out of its light cone.

The computations needed to manage a light require a considerable amount of CPU time. It is thus crucial to choose the right kind of light related to the scope of a scene and to avoid putting lights on every object by using a *scope node* and thus saving computational performance. Every light has a color determined by the RGB components and has an intensity value, but the exact effect of light hitting a surface is also function of that surface's material.

4.9 Meshes and Sprites

A mesh object is a node in a scene graph that represents a three-dimensional3D object specified by a set of polygons. The object itself is made of several submeshes, each having its own appearance, as shown in Figure 4.6.

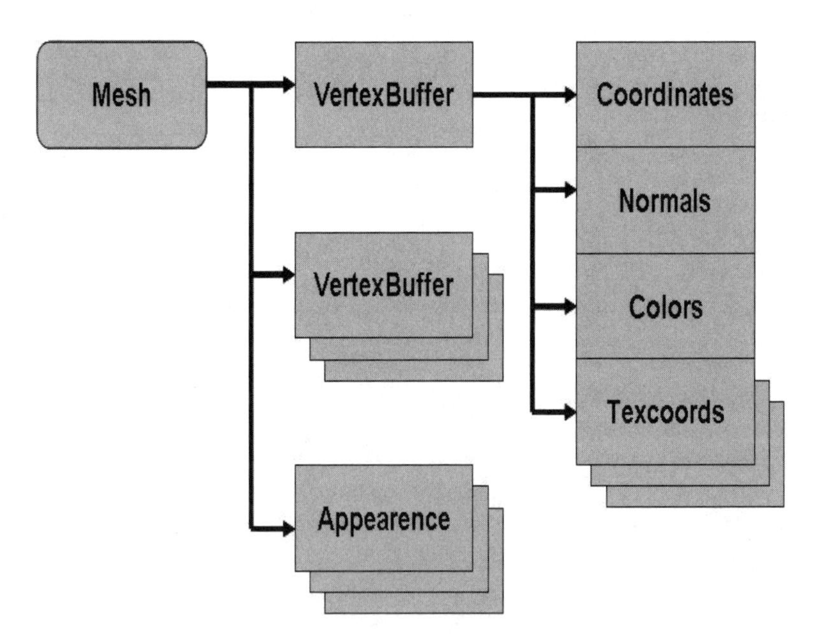

Fig. 4.6. A mesh object structure.

A submesh is an array of *triangle strips* defined by an *IndexBuffer* object. *Triangle strips* are made by indexing vertex coordinates and other attributes of a *VertexBuffer* associated with an *IndexBuffer*. *VertexBuffer* contains information about vertex positions, normals, and texture coordinates. Each submesh in a mesh shares the same *VertexBuffer*.

The components of an appearance object are:

- *Material*, which defines the colors to be used in lighting computations.
- *CompositingMode*, which allows per-pixel composition attributes such as transparencies and z-buffer.
- *PolygonMode*, which contains attributes at the polygonal level including settings for face visibility (back and front) and perspective corrections.
- *Fog*, which contains all attributes for setting a fog effect.
- *Texture2D*, which incorporates all 2D images and attributes for specifying how an image can match the related submeshes.

The *mesh* class also includes two subclasses used for managing dynamic meshes, which can change their shapes according to certain parameters: *MorphingMesh* and *SkinnedMesh*.

An object of *MorphingMesh* type is equivalent to an ordinary mesh, except that its vertices are drawn and computed as a weighted linear combination. It is a combination of a *VertexBuffer* and *VertexBuffers*, which are targets of the morphing operation. All target *VertexBuffers*, also called *morph targets*, include the same properties: the same number of vertices for each array, the same number of components per vertices, and the same component size.

By denoting a base mesh by B, morph targets by T_i, and weights for each morph target by w_i, a resulting mesh can be represented by the following equation:

$$R = B + \sum_i w_i(T_i - B)$$

Morphing can be computed on every vertex attribute:

- Vertex positions
- Colors
- Normals
- Texture coordinates

The *SkinnedMesh* class represents a skeleton animated polygonal mesh. In contrast with a normal mesh class, it includes a skeleton structure. The skeleton is built by means of a hierarchical structure, by using scene graph nodes. Each node belonging to a skeleton represents a *bone*, which is a transformation. Each vertex can be linked to one or more skeleton bones. In this way a mesh is extended and linked to a structure that can manage it.

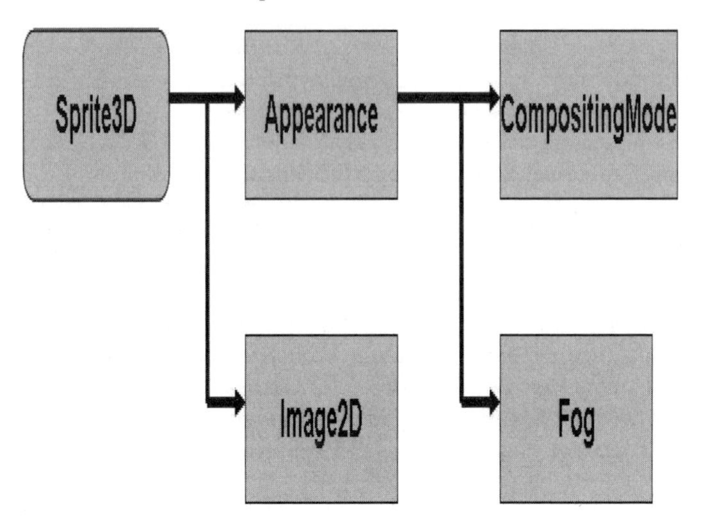

Fig. 4.7. A Sprite3D object structure.

The *Sprite3D* class represents a 2D image with a position in three-dimensional space. The structure of a Sprite3D object is shown in Figure 4.7.

Images are stored in *Image2D* objects. Their *appearance* contains attributes for fog and composite effects. There are two modalities for appearance:

- *Scaled mode*, in which the width and height of a sprite on the screen are computed, as it is a rectangle with one unit thick and based on the XY plane centered in its local coordinate system origin.
- *Unscaled mode*, in which the width and height of a sprite are measured in pixels and are equal to a rectangle defined by setting its size.

4.10 Animations

Each object extended from a basic *Object3D* class can be animated. The most relevant classes for managing animations are:

- *KeyFrameSequence*
- *AnimationController*
- *AnimationTrack*

KeyFrameSequence contains all animation data as a time sequence of values called *keyframes*. A *keyframe* represents a value of an attribute at a certain instant of time. It contains a vector of components, specified by its constructor, which has the same size for each *keyframe* in a sequence. Since *keyframe* values can be distant in time, interpolation functions are provided to manage them.

A *KeyFrameSequence* object can be associated with different *animation targets* by using an *AnimationTrack* class. It associates a *KeyFrameSequence* with an *AnimationController* object and a property that can be animated. This kind of property consists of a scalar value or a variable vector that can be updated by an animation system. An example of a property that can be animated is the orientation of a node. Animated properties are identified by a symbolic constant, and sometimes they're related only to a restricted class of values, like the *SHINESS* property of a *Material* object.

Classes derived from *Object3D* include one or more animated properties. An *Object3D* with animated properties is called an animated object. Each animated property of an animated object is an *animation target*. Each animated object can include references to zero or more *AnimationTracks*, which are activated by their related *AnimationControllers*.

An *AnimationController* manages the position and speed of an animation sequence. An animation sequence can be defined as a set of *AnimationTracks* managed by a single *AnimationController*. Each *AnimationTrack* contains all the data needed to manage an animated property on an animated object. By using an *AnimationController*, operations like pausing, stopping, playing, and speeding-up an animation sequence are available.

Figure 4.8 shows the classes related to the animation process.

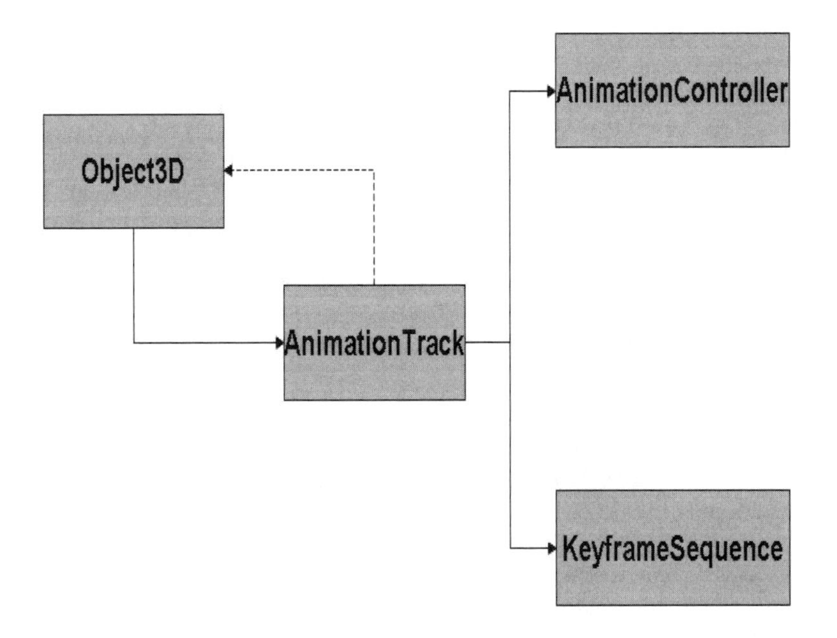

Fig. 4.8. Classes related to the animation process.

4.11 Ray Intersections

The *RayIntersection* class represents an infinite line starting from an origin (in a coordinate reference system) and pointing in a fixed direction. It is used for storing references to all *Mesh* or *Sprite3D* objects intersected by this line. Not only intersections but also distances between this line and intersected objects are stored. The *RayIntersection* object is created at run time and cannot be loaded by the *Loader()* class. It is used in conjunction with the *pick()* method of a *Group* class. This method returns information on this first object in a group intersected by the line passed as a parameter to this method. Information on intersected objects is then returned by the *RayIntersection* object. This class is used to manage collisions among objects or to simulate, for example, a gun-shot hitting a target placed at a fixed distance.

4.12 Building an M3G Demo

In this section we explore how to use some high-level classes provided by M3G API to create a simple demo program with a car model moving on the screen and avoiding obstacles by a collision detection method.

First we need to install a *Sun Java Wireless Toolkit*, and create a new project by using a *Ktoolbar* interface. This will help in understanding applications of the described concepts and classes.

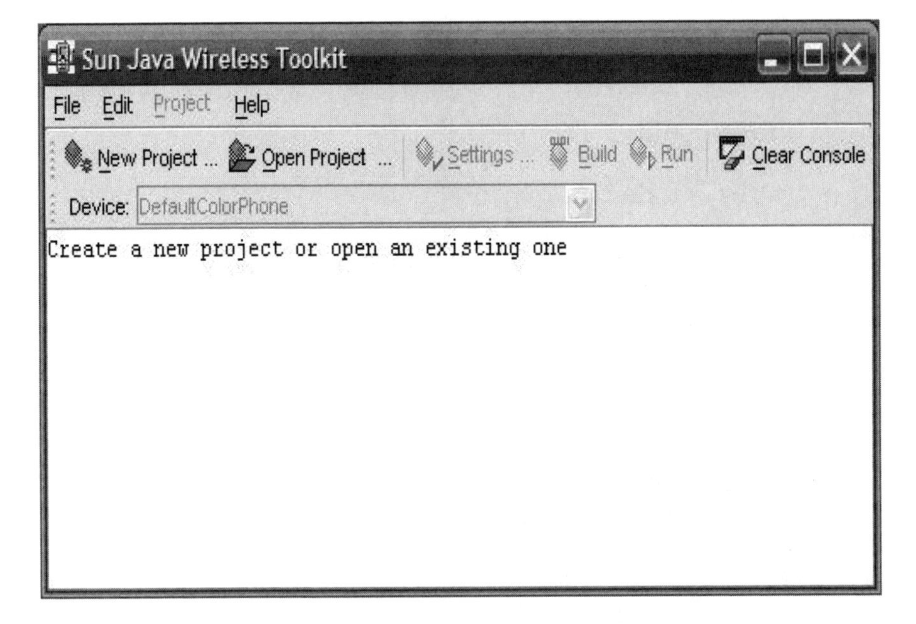

Fig. 4.9. A snapshot of a wireless toolkit interface.

After defining both the project and MIDlet name, we need to choose a CLDC 1.1 configuration, which supports float data type, and choose Mobile 3D Graphics for J2ME (jsr 184) as additional API.

Recall that Figure 4.3 showed the life cycle of a MIDlet and methods used for changing MIDlet status, which will be used below.

The example of source code for a MIDlet includes not only a declaration of these methods, but also some functions for timing animations.

```
import javax.microedition.midlet.*;

import javax.microedition.lcdui.*;

import java.util.*;

import java.io.*;

public class CarDemo extends MIDlet {
    private static final int PERIOD = 50;   // in ms
    private Timer timer;
    private CarDemoCanvas canvas = null;
```

```
private Display display;

public void pauseApp() {}

public void destroyApp(boolean b){}

public Display getDisplay()
{
    return display;
}

public void startApp() {

// check whether m3g is supported or not
String version =
System.getProperty("microedition.m3g.version");
if (version == null) {
        finishGame();
    }
else {
        display = Display.getDisplay(this);

        canvas = new CarDemoCanvas(this);
        timer = new Timer();

        display.setCurrent(canvas);
        timer.schedule( new AnimTimer(),0,PERIOD);
    }
}
public void finishGame() {
    timer.cancel();     // stop the timer
    notifyDestroyed();
}
```

Inside the *StartApp()* method a *microedition* version property is checked to indicate if the device supports additional *m3g* API. In case of a positive answer, the *CarDemoCanvas* class is initialized as the core class of our demo application.

In the same source code must be inserted a class called *Animtimer* that is used for managing animations. This class contains a *run* method, which is in charge of updating animations. To set the timing for animations, a *Timer* class has been used.

```
// Class for managing timing
  class AnimTimer extends TimerTask
  {
    public void run()
    { if(canvas != null)
        canvas.update();
    }
  }
```

4.12.1 DemoCarCanvas Class

The *DemoCarCanvas* class extends the *Canvas* class and implements a *CommandListener* for managing the code to be executed in response to the *EXIT* and *BACK* events generated by the device user's interface.

The following snippet of code is a class constructor.

```
public CarDemoCanvas(CarDemo carDemo){

        this.carDemo = carDemo;

        exitCmd = new Command("Exit",Command.EXIT,0);
        addCommand(exitCmd);
        setCommandListener(this);

        g3d = Graphics3D.getInstance();

        width=getWidth();
        height=getHeight();

        scene = new World();

        createScene();

        // start the animation
        nextTimeToAnimate = scene.animate(appTime);

    }
```

First a reference to the MIDlet CarDemo is stored, and then the *EXIT* command is set to allow users to close the application by pressing a device key.

After the *Graphics3D* object is instantiated, it represents a 3D graphics context and provides a method for scene drawing, called *render*.

The device screen height and width are set, and after the *scene* object is created (the type of this object is *World*), it will contain all the three-dimensional scene objects (lights, cameras, and meshes).

Finally the *createScene* method is invoked for creating and setting all objects included in the scene.

```
private void createScene() {
    createCar();
    createCamera();
    createLight();
    createBackground();
    createFloor();
    createCone();
}
```

In *CreateScene*, many methods are called, one for each object included in the three-dimensional scene.

In the scene we include:

- One camera (normCamera)
- Two lights (light and light2)
- One background (background)
- Three meshes for visualizing a car, some cones (alias the obstacles), and a floor

To load the mesh models in a three-dimensional scene, a technique has been developed by *Andrew Davison* [44] which converts a *wavefront OBJ* model into a Java class containing M3G code. We thus generated three different classes, Car, Floor, and Cone, including all visualization code for these three models.

A *createCamera* class sets up a simple camera, and we will use some transformations on this camera (mainly two 90-degree rotations with respect to the x and y axis) for visualizing the scene from the right perspective. To have a better perspective, there is also a *setOrientation* method, which slides the scene down a little bit.

Finally a camera is added to the *scene* object and set as active.

```
private void createCamera() {

    float aspectRatio = (float)width/ (float)height;
```

```
    // normCamera
    normCamera = new Camera();
    normCamera.setPerspective(60.f, aspectRatio, 1.0f,
    100000.f);

    // camera transformations
    Transform normCameraTransform = new Transform();
    normCameraTransform.postRotate(90, 1f, 0f, 0f);
    normCameraTransform.postRotate(90, 0f, 1f, 0f);

    normCameraTransform.postTranslate(0f, 0f, 200.0f);
    normCamera.setTransform(normCameraTransform);

    // angles downward slightly
    normCamera.setOrientation(-50.0f, 0f, 1f, 0f);

    scene.addChild(normCamera);
    scene.setActiveCamera(normCamera);
}
```

Code that manages the lights also sets and uses the methods and properties of M3G API as shown below.

```
private void createLight() {

    // 1 omni light (ahead)
    Light light = new Light();
    light.setColor(0xffffff);
      light.setIntensity(1.0f);
    light.setMode(Light.OMNI);
    light.setTranslation(0,0,100);

    // 1 omni light (behind)
    Light light2 = new Light();
    light.setMode(Light.OMNI);
    light2.setIntensity(1.0f);
    light2.setTranslation(100,100,100);

    scene.addChild(light);
    scene.addChild(light2);
}
```

Very similar to the method above is the *createBackground* method coded as follows:

```
private void createBackground() {

    Background background = new Background();

    background.setColor(0x004080C0);

    scene.setBackground(background);
}
```

This method creates a background object, setting it to a light blue color. The color format is managed by M3G as *0xAARRGGBB*, where:

- A stands for the alpha channel
- R stands for the red channel
- G stands for the green channel
- B stands for the blue channel

The remaining methods, *createCar*, *createCone*, and *createFloor*, use their respective classes: Cone, Floor, and Car.

```
private void createFloor() {

    Image2D floorIm = loadImage("/piano.png");
    Plane plane = new Plane( floorIm, 1000,1000);
    scene.addChild( plane.getPlaneMesh() );

}


private void createCone() {

    Cone cone  = new Cone();
    Cone cone1 = new Cone();
    cone.getMesh().setTranslation(100,130,30f);
    cone1.getMesh().setTranslation(-150,-100,30f);
    scene.addChild(cone.getMesh());
    scene.addChild(cone1.getMesh());
}
```

The *createFloor* and *createCone* classes, respectively, build a plane and two cones (obstacles), similar to street cones positioned in the scene.

The *Plane* class creates a 2D plane, and its constructor takes three parameters as input:

- An *Image2D* object for managing textures
- An integer representing x axis elongation
- An integer representing y axis elongation

All images are loaded by the *loadImage* method, which performs some consistency checks before returning an *Image2D* object as result.

```
private Image2D loadImage(String fn) {
        Image2D im = null;
        try
{
                im = (Image2D)Loader.load(fn)[0];
        }
        catch (Exception e)
        {
System.out.println("Cannot make image from " + fn); }
        return im;
}
```

The *createCar* method does not contain a *Car* class constructor, since a *car* object is created during the initialization phase; it contains instead methods for managing collisions among car and other objects (street cones) in the scene:

- The *setScene* method takes a World object pointer to use it with the *Car* class.
- The *setPickingEnable* method enables/disables a car mesh during collision computations.

```
private void createCar() {

    // Disable collision detection for car mesh
    car.getMesh().setPickingEnable(false);

    car.setScene(scene);

    carGroup = car.getCarGroup();

    scene.addChild(carGroup);
}
```

The *update* method is invoked for updating the car position on the screen after checking for collisions by using the *updateCar* method of *Car* class. Once a car is positioned, the screen is repainted by the *repaint* method.

```
public void update() {
    appTime++;
    if (appTime >= nextTimeToAnimate) {
        nextTimeToAnimate = scene.animate(appTime)
        + appTime;
        System.out.println("nextTimeToAnimate: "
        + nextTimeToAnimate);
    }

    car.updateCar();

    posCar = car.getPosition();
    normCamera.setTranslation(posCar[0], posCar[1],
                              posCar[2]);

    repaint();
}
```

The *paint* method is in charge of drawing the final three-dimensional scene. As already mentioned, the only method provided by the M3G standard for drawing is the *render* method; it can be called after linking a graphics context to a canvas.

We also display on the screen the car speed (top left side of the screen).

```
protected void paint(Graphics g) {
    g3d.bindTarget(g);

        g3d.render(scene);

    g3d.releaseTarget();

    g.drawString( "Speed: " + car.getSpeed(),5,5,
    Graphics.TOP|Graphics.LEFT);

}
```

The last two methods of the *CarDemoCanvas* class are used for managing the keyboard.

```
protected void keyPressed(int keyCode)  {
    int gameAction = getGameAction(keyCode);

    car.pressedKey(gameAction);
```

```
}

protected void keyReleased(int keyCode) {
    int gameAction = getGameAction(keyCode);

    car.releasedKey(gameAction);
}
```

Both methods check which key has been pressed, store it in a *gameAction* variable, and pass it to *Car* class methods for updating the car position.

4.12.2 Car Class

Car class not only contains code for visualizing a car model (i.e., *Floor* and *Clone* classes) but also has methods for animating cars, and manages (by means of the *RayIntersect* class) a basic collision detection algorithm.

The car class constructor takes an *Image2D* parameter for textures. It also is in charge of building a car model and managing a group of transformations (*trans*) for positioning a car in the scene. The model is then linked to a group; in this way if we modify a transformation each of the children nodes is affected by this change.

```
public Car(Image2D img) {
    this.scene=scene;

    model = makeModel(img);

    transGroup = new Group();
    trans.postTranslate(X_POS, Y_POS, Z_POS);
    transGroup.setTransform(trans);
    transGroup.addChild(model);
}
```

The *storePosition* method extracts the current car position from the transformations group.

```
private void storePosition()
// extract the current (x,y,z) position from transGroup
{
    transGroup.getCompositeTransform(trans);

    trans.get(transMat);
    xCoord = transMat[3];
```

```
    yCoord = transMat[7];
    zCoord = transMat[11];
}
```

The *transMat* object represents a 4×4 float matrix:

```
private float[] transMat = new float[16];
```

The methods used for managing position and direction of the car are:

- *updateMove*, which takes a transformation parameter (*trans*), performs all checks for collision detection, and moves the car according to a space attribute computed by the current speed of the car.
- *updateRotation*, which by means of key pressed (left and right arrow keys) rotates the car (using a *Transform* object called *rotTrans*).

Both these methods are invoked by the *updateCar* method, which manages the following:

- Increasing speed until the up arrow key is released, and decreasing speed when holding down the arrow key (or releasing both keys)
- Executing the *updateMove* method
- Executing the *updateRotation* method only if the left or right arrow keys are pressed

```
public void updateCar() {

        if (upPressed) {
            if(speed<MAX_SPEED) speed+=2f;
        }

        if(downPressed) {
            if(speed>0.0f) speed-=4f;
            if(speed<0.0f) speed=0.0f;
        }

        if(!upPressed && !downPressed) {
            if(speed>0.0f) speed-=2f;
            if(speed<0.0f) speed=0.0f;
        }

        updateMove();

        if (leftPressed || rightPressed)
```

```
            updateRotation();
        else if(!leftPressed && !rightPressed)
                angle=2.0f;
}
```

All necessary attributes for managing key pressing (*Boolean*) are updated by the *pressedKey* and *releasedKey* methods:

```
public void pressedKey(int gameAction) {
    switch(gameAction) {
            case Canvas.UP: upPressed = true;    break;
            case Canvas.DOWN: downPressed = true;  break;
            case Canvas.LEFT: leftPressed = true;  break;
            case Canvas.RIGHT: rightPressed = true; break;
            default: break;
    }
}
```

```
public void releasedKey(int gameAction) {
    switch(gameAction) {
            case Canvas.UP: upPressed = false;  break;
            case Canvas.DOWN: downPressed = false; break;
            case Canvas.LEFT: leftPressed = false; break;
            case Canvas.RIGHT: rightPressed = false;break;
            default: break;
    }
}
```

We now describe the *updateRotation* method as coded below.

```
private void updateRotation() {
    if(angle<MAX_ANGLE) angle+=1.0f;

    if (leftPressed) {    // rotate left around
                          // the z-axis
            rotTrans.postRotate(angle, 0, 0, 1.0f);
            zAngle += angle;
    }
    else {   // rotate right around the z-axis
            rotTrans.postRotate(-angle, 0, 0, 1.0f);
            zAngle -= angle;
```

```
    }

    // angle values are modulo 360 degrees
    if (zAngle >= 360.0f)
            zAngle -= 360.0f;
    else if (zAngle <= -360.0f)
            zAngle += 360.0f;

    // apply the z-axis rotation to transGroup
    storePosition();
    trans.setIdentity();
    trans.postTranslate(xCoord, yCoord, zCoord);
    trans.postRotate(zAngle, 0, 0f, 1f);
    transGroup.setTransform(trans);
}
```

Rotation takes place on the axis by changing the *zAngle* attribute, which is increased or decreased by pressing the appropriate key. It is important to store the rotation status in the *Transform rotTrans* object for keeping the information useful for the car direction vector.

The *getDirection* method computes the direction of the car.

```
public float[] getDirection() {

    // zVec contains the initial direction of the car
    float[] zVec = {-1, 0, 0, 0};

    // the exact direction is given after
    // computing the applied rotations

    rotTrans.transform(zVec);

    return new float[] { zVec[0], zVec[1], zVec[2] };
}
```

The car direction is obtained by computing all the applied rotations. Car direction is used for computing collision detection in the *updateMove* method.

Collisions are managed by using the *RayIntersect* class provided by the M3G standard. A *RayIntersect* object is set by the *pick* method, which is part of the *Group* objects. *RayIntersection* stores a pointer to the intersected *Mesh* or *Sprite3D*, and to all the relevant information about the intersection point.

The *pick* method first takes *Mesh* or *Sprite3D* into the group and enables it for picking, which is intersected by a pick ray passed as a parameter (a ray is a line in our case).

The following is the code for the *updateMove* method. It is self-explanatory, as it includes comments:

```
private void updateMove() {

    transGroup.getTransform(trans);

    // computing space from speed
    space=speed*0.5f;

    // check collisions
    RayIntersection ri = new RayIntersection();

    // car direction updating
    dir=getDirection();

    // check whether there is something in front
    // (Mesh or Sprite3D) excluding the car itself
    // and the floor

    if (scene.pick(-1, xCoord, yCoord, zCoord,
    dir[0], dir[1], dir[2], ri)) {

        // object distance
        float distance = ri.getDistance();

        // 38 is cone size

        if(distance>38.0f + space) {
                // move
                trans.postTranslate(-space, 0, 0);
                transGroup.setTransform(trans);
        }
        else {
            // stop the car
            speed=0.0f;
            return;
        }
    }
```

```
    // move
    trans.postTranslate(-space, 0, 0);
    transGroup.setTransform(trans);

}
```

Figure 4.10 shows a snapshot of the application described with coding examples, and includes many of the API described in the chapter.

Fig. 4.10. DemoCar screen shot.

4.13 Summary

This chapter introduced M3G and the Java Mobile 3D Graphics library, and described how an application could be developed for mobile devices supporting this standard.

We described also the frameworks (CLDC/MIDP) used by Java for managing mobile devices and applications. M3G is consider an extension of these libraries and thus it is included in their development process. We also discussed the modalities of M3G, *Immediate* and *Retained* mode, explaining when and how to choose between the two. We then described elements of the M3G scene graph, which is a hierarchical structure used by this library for representing and managing a 3D scene.

We finally provided a comprehensive example, called *CarDemo*, including all the concepts, elements, and API that clarify the functionalities.

5

Direct3D®Mobile

Direct3D®Mobile (D3DM) is a Microsoft™-developed API that provides 3D support for mobile devices based on Microsoft Windows®OS. It is derived directly from DirectX®API already included in the desktop versions of MS Windows; moreover, it is optimized to match mobile devices' requirements. The main source of information concerning D3DM is the *Microsoft Developer Network* library (MSDN) [45]. This chapter discusses D3DM libraries and describes the architecture of these API; we consider that D3DM API, compared to OpenGL ES and M3G, suffer from their portability; as it can be used only with MS Windows OS.

5.1 Architecture

D3DM is implemented with component object model (COM) interfaces and objects; COM is an object-oriented protocol used by Microsoft products that is quite powerful and extensible. All of the most used and current programming techniques in the MS Windows environment, like .NET®and COM+®, are based on COM libraries. From a technical viewpoint the COM architecture is very extensible since it allows a combination of software components even during run-time execution. That is like having off-the-shelf facilities for programming.

COM technology provides primitives for developing reusable software modules; these modules expose functions that modules can invoke by using programming interfaces. All objects included in COM architecture can be developed by using two different approaches:

- They can be stored in dynamic linking libraries (DLL).
- Or they can be linked directly in an executable file (EXE).

After describing very generally the COM technology, we focus on the D3DM approach (based on COM). The first object a D3DM application can create and interact with is the *Direct3DMobile* object. When an application written to use this technology is executed, it must obtain a pointer to the *IDirect3DMobile* interface (an interface for the *Direct3DMobile* object) in order to have access to all its functionalities.

The following code snippet shows how to invoke a *Direct3DMobileCreate* function in order to get a pointer to a *IDirect3DMobile* interface:

```
LPDIRECT3DMOBILE g_pD3Dm = NULL;

    if ( NULL == (g_pD3Dm =
    Direct3DMobileCreate(D3DM_SDK_VERSION)))
        return e_FAIL;
```

The API functionalities are grouped together in a set of interfaces that provide access to standard COM methods; not all of these interfaces are allowed to create objects or invoke other interfaces by themselves.

The three main aspects of Direct3D Mobile are as follows:

- Graphic models described by geometric primitives
- Geometric primitives parameters and drawing options encapsulated in a finite state machine called the *rendering pipeline*
- Output provided by a frame-based graphic model

Basic primitives included in the D3DM library are points, lines, and triangles. They are all described in term of vertices in a three-dimensional Cartesian space. Vertices data are all loaded in particular structures called *vertex buffers*, described in Chapter 3. The D3DM library provides applications with methods for creating *vertex buffers* and for loading and describing vertex data.

The *Rendering pipeline* in D3DM is designed as a finite-state machine used for managing geometric primitives display. Inputs to a finite-state machine are geometric primitive descriptions, while output is represented by different color values of pixels contained in a frame buffer. The Direct3D Mobile library provides a set of options and parameters, called states, for managing the rendering phase. The possible statuses that can be used by a developer are provided by a set of definitions and properties (called *capabilities*) included in the D3DM software drivers.

There are two methods for a software driver to represent the properties it supports:

- A low-level method uses a standard set of property bits. Applications can query these bits to get information about the graphics level and the features supported by that particular driver.
- Another method is based on system profiles built on top of property bits. Each profile represents a specific set of bits packaged for different categories of mobile devices. In this way applications developed for a certain profile

work with drivers supporting that specific profile. This approach simplifies the development of mobile graphics applications because it directly uses profiles with no need for an application to query the property bits.

D3D mobile uses a display model based on frames. This means that an application notifies the API that it is ready to display by invoking the *IDirect3DMobileDevice::BeginScene* method. While that application defines a scene, API fill a *command buffer*, which contains all the commands (with primitives) for drawing that scene. When an application finishes drawing a scene, it invokes the *IDirect3DMobileDevice::EndScene* method. After these steps the scene is enqueued for the rendering phase but it is still not passed to a driver. The application should explicitly invoke a *IDirect3DMobileDevice::Present* method in order to display scenes on a mobile device.

5.2 Rendering Pipeline

The *rendering pipeline* defines how to manage and process input data, usually vertices, by transforming them in pixel color values. This computational process joins three-dimensional data models to a graphic display. It offers a set of predefined options that describe how input data have to be processed. These options can be applied in different ways, even if the number of available options is limited. Basically all rendering processes can be subdivided into four phases:

- Transformation
- Lighting
- Rasterization
- Per-pixel operations

During the first phase, *Transformation*, vertices are numerically processed to generate physical coordinates on the screen for points, lines, and triangles. The D3DM library simulates lighting effects by using mainly vertices. When *Lighting* is enabled, light parameters are used for computing diffuse colors and specular values for each pixel in the scene. These values are passed to a *rasterizer*, which is in charge of interpolating values for each pixel in the scene. In the third phase, *rasterization*, the screen space coordinates for various primitives are involved in determining the exact set of pixels composing a specific primitive. The scope of this phase consists of finding vertices that are part of a geometric primitive for generating pixels that will display the primitives on screen. The components of a vertex, like color and texture data, are used for representing the attributes of a geometric primitive. After computing the output values for each pixel, some final operation should be performed in order to get a rendered scene, *Per-Pixel operations*. These operations include many tests for example, for determining the visibility of a pixel and its level of transparency in a scene. Some of these operations result in discarding a specific pixel from a scene (for example, if it is occluded by other pixels), and in that case the process will continue on the next pixel belonging to the specific primitives.

5.3 Primitive Types

Many primitive types are supported by D3DM, and thus could be supported by an application written for this platform. These types are passed as parameters to *IDirect3DMobileDevice::DrawPrimitive* and *IDirect3DMobileDevice::DrawIndexedPrimitive* in order to be drawn.

The following table lists some supported types:

Primitive types	D3DMPRIMITIVETYPE value
Point List	D3DMPT_POINTLIST
Line List	D3DMPT_LINELIST
Line Strip	D3DMPT_LINESTRIP
Triangle List	D3DMPT_TRIANGLELIST
Triangle Strip	D3DMPT_TRIANGLESTRIP
Triangle Fan	D3DMPT_TRIANGLEFAN

The vertices used by each primitive are read by an active *vertex buffer*. The *IDirect3DMobileDevice::DrawPrimitive* method takes as a parameter a pointer to the first vertex, from which it reads the rest of the data in sequence, and a number of primitives (to be drawn). If there aren't enough available vertices in the *vertex buffer* for drawing a primitive, it is ignored and the process continues with successive commands. The *IDirect3DMobileDevice::DrawIndexedPrimitive* method uses the same primitive types first method but the vertices are managed differently. Instead of loading the vertices directly from a vertex buffer in sequence, the indices (pointing to vertices) are loaded from an *index buffer*. Each index is a pointer to a corresponding vertex stored in the *vertex buffer*. In this way each vertex stored in a single position in the vertex buffer can be used many times without wasting memory by replicating the same data. Vertices could include different kinds of data, but have a position vector for locating the vertices in coordinate space. A set of different data associated with a vertex (for instance, a color for light diffusion and a color for specular diffusion) is defined by a flexible format (Flexible Vertex Format, *FVF*). Vertex structures, stored in a *vertex buffer*, are created by invoking the *IDirect3DMobileDevice::CreateVertexBuffer* method. One parameter of this method is an *FVF* value. This means that all vertices included in a *vertex buffer* must have the same data format.

We will now show a typical section of D3DM code for creating a *vertex buffer*. This will help readers to better understand the programming statement logic included in the D3DM library, which is, as already mentioned, close to the Microsoft DirectX programming logic; thus a pseudo-code snippet will look like this:

```
protected VertexBuffer CreateVBuff(Device dev)
 {
   dev.VertexFormat = setVertextFormat(vertext.Format);

   vertices[] = new vertices[3];

   vertices[0] = new vertices(x0, y0, z0, 1,
   Color0.ToArgb());
   vertices[1] = new vertices(x1, y1, z1, 1,
   Color1.ToArgb());
   vertices[2] = new vertices(x2, y2, z2, 1,
   Color2.ToArgb());

   VertexBuffer buffer = new VertexBuffer(
   typeof(vertices), vertices.Length, dev, 0,
   vertices.Format, Pool.Default);

   GraphicsStream stream = buffer.Lock(0, 0, 0);
   stream.Write(vertices);
   buffer.Unlock();
   return buffer;
 }
```

Since vertices in D3DM are described by an FVF, the first lines set the vertex format as a global setting. This means that a device is now expecting every vertex to have transformed coordinates and to contain color information.

After the format statement we create a vertex array for storing vertices information.

We store information in this array by using five parameters: x, y, and z position, a w value, and an integer representing a color. Let's ignore z for now and recall that we are working in transformed coordinates (screen coordinates) and so z does not have to be managed for it. The w could also be ignored for this first example; in fact, it is set to a constant value 1. Color is clearly referred to a vertex, although D3DM makes use of the *ToArgb()* method to convert it to an integer.

By defining three vertices we are basically creating a triangle, but we still need to create a *vertex buffer* itself.

Note that a *VertexBuffer* can contain one type of vertex. We define what type it is with a first argument, how many vertices it includes with a second argument, and a device with the third. We specify options with a fourth argument (set to 0 for now), defining the format of the vertices with fifth, and ask for default pooling with the final argument. Pooling is an advanced topic, and we don't need to deal with it in this simple example.

Finally, we get a reference to a *GraphicsStream* by locking *VertexBuffer*, and then call its method for writing three vertices into the *VertexBuffer*. Then we need to unlock *VertexBuffer*. Lock and unlock are important because they tell D3DM that we are updating *VertexBuffer* and it can't, for instance, try to access information while we are doing so.

With *VertexBuffer* created, we can now concentrate on rendering, as shown in the next snippet of code.

```
protected void Render()
{
    dev.Clear(Color);

    dev.BeginScene();

    dev.SetStreamSource(0, vertices, 0);

    dev.DrawPrimitives(PrimitiveType.TriangleList, 0, 1);

    dev.EndScene();

    device.Present();
}
```

The first two lines simply clear a device context with a background color and set a scene starting point.

SetStreamSource informs a device to render by taking vertex data from that *vertex buffer*. In fact, there might exist several *vertex buffers*, for instance one for each object that we want to display; thus it is important that we set a stream source before each rendering call, otherwise the system will display the wrong shapes.

DrawPrimitives takes as the first argument things to draw. There are basically six choices: *PointList*, *LineList*, *LineStrip*, *TriangleList*, *TriangleStrip*, and *TriangleFan*.

Since we are drawing a triangle, we have three choices: *TriangleList*, *TriangleStrip*, and *TriangleFan*. *TriangleList* is the simplest because it specifies to D3DM that we are going to handsets of three vertices in a *vertex buffer*, and for each triplet it draws a corresponding triangle. We won't mention again the triangle fans and strips since they work the same as in OpenGL ES API.

A second parameter is an offset of a *vertex buffer*. In our case, the offset is zero, since we want to start at the beginning. But if we have to manage hundreds of vertices describing many triangles, we might want to draw only a small portion of them. In that case we need an index for the first triangle to display.

The third parameter controls the number of primitives that will be drawn. For instance, specifying 2 in our code snippet would cause an error, since the system would try to draw two triangles, which for a TriangleList would need six points, while instead our *vertex buffer* only has three (Figure A.2).

5.4 Transformations

Transformations are performed, as in standard OpenGL, by using matrices called *transformation matrices*. These matrices can be defined using floating point 32-bit IEEE standard values, or using fixed point values of 16.16 type (see Chapter 2, section 2.3.1).

The default value for transformation matrices is an identity matrix. There are three types of transformations:

- World transformations
- View transformations
- Projection transformations

The first transformation (*World*) transforms coordinates from a model space, where vertices are defined with respect to a local model coordinates origin, to global space, where vertices are defined with respect to a common coordinate origin of all objects included in a scene. Figure 5.2 shows the relationships between the global coordinates and the local model coordinates.

The second transformation (*View*), localizes an observer viewpoint in world space; thus it transforms the vertex coordinates in the camera space coordinates. In this space, a camera is located in the origin of the space, directed toward the positive z-axis.

The third transformation (*Projection*) is related to the camera's internal controls and it is a bit more sophisticated. In this phase a view volume

Fig. 5.1. In the rendered triangle the color of each vertex is blended smoothly across triangle surface. We have set *Color0.ToArgb()* as red, *Color1.ToArgb()* as green, and *Color1.ToArgb()* as blue.

is defined, a three-dimensional volume represented by (using perspective) a pyramid truncated by the front and back planes (clipping planes).

In the previous section, we discussed how to render a scene using a *vertex buffer*. During the rendering phase we used transformed coordinates.

Rendering a scene involves many steps. The most effective way of modeling an object differs between each of these steps. For example, if we consider a cube, when specifying vertices that compose that cube, it is easier to set one of its corner as being located "in origin", and other corners being at coordinates such as $(0, 0, d)$, $(0, d, 0)$, $(d, d, 0)$, etc.

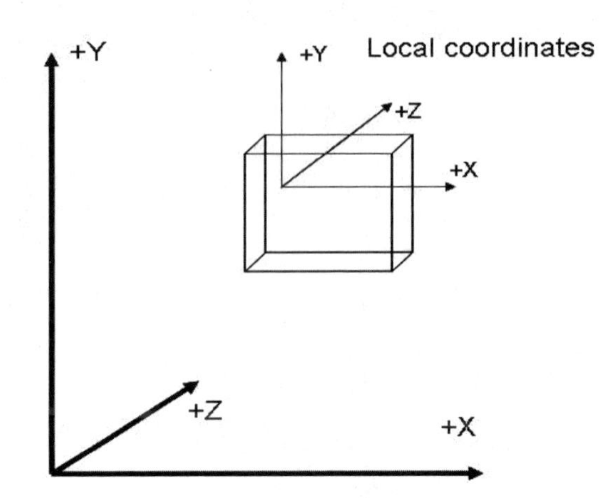

Fig. 5.2. Relationships between local and global coordinates.

A coordinate system appropriate for a given object is called a *local coordinate system*, *local space*, or *object space.*

If we have a complex scene, like hotel room with many objects inside it, such as a: bed, desk, and lamps, we need a coordinate system that is different from a local coordinate system. We will set up a coordinate system that has its origin in a corner of that room, and whose axes are lined up with the walls and floor. This coordinate system is called *world coordinates* or *world space.* The idea is that it is a coordinate system that expresses the positions of things in the world. All objects in a scene have their own set of local coordinates, but they share world coordinates, as already seen in the first chapter.

To transform coordinates from one coordinate space into another, we need to manage matrix operations. If we pick our matrix elements carefully, multiplying that matrix with a set of coordinates in one coordinate space turns them into equivalent coordinates in another space. D3DM refers to this as a transformation, and there are several important ones. In that sense, a matrix is a transformation. For instance, a transformation that helps us locate our object in world space (turns the objects' local coordinates into world coordinates) is a world transformation.

A matrix structure, included in D3DM library, has lots of methods that free us from having to execute math computations for finding the appropriate matrix coefficients.

For example, we can have a matrix that represents a k fraction of 180-degree rotation around the x axis:

```
Matrix worldTxfm = Matrix.RotationX(Math.PI / k);
```

For instance if $k = 2$, the rotation will be of degrees, since the angle is specified in radians.

There are mainly two kinds of transformation in D3DM: view transformations and projection transformations. Lets consider the view transformation first.

While object space is a coordinate system that is useful for objects in a scene, certain operations needed for rendering are more easily represented in view space. There is also a *camera space*, because space is defined as having its origin at a camera (or eye) point. That is, its origin is co-located with a viewer. In addition, to set an origin at a camera point, the view space z axis points to the same direction as the camera, and the y axis points up. We can have a matrix that represents transformations necessary to change world coordinates into view space coordinates using a matrix method called *LookAtLH*.

```
Matrix viewTransform = Matrix.LookAtLH(
   new Vector3(x,y,z),   ...
 );
```

LookAtLH takes three arguments: camera position, position to point at, and a direction for "up". These coordinates are specified by a *Vector3* structure in world coordinates.

View space is useful for graphical operations that have to interact with a camera (such as z-buffering) but its not very useful for displaying pixels on the screen. To turn three-dimensional objects' view space into two-dimensional display space, we can use a projection transformation. Matrix structure has a method also for projection transform:

```
Matrix projectionTransform = Matrix.PerspectiveFovLH(
   field_of_view, aspect_ratio, front_plane, back_plane
 );
```

PerspectiveFovLH takes four arguments: a field of view, an aspect ratio, and distances that define clipping planes.

For field of view larger numbers will display more details of world space on the screen, but will probably end in distortions. Smaller numbers mean less distortion, but less world will be displayed.

The aspect ratio indicates the $\frac{width}{height}$ ratio of a displayed window. Note that regardless of the aspect ratio, D3DM will always change images to fit the actual windows.

The last two arguments specify the clipping planes. We already described this concept in a previous chapter, but recall that it can be defined as a plane that specifies the front or back of a scene, that is a portion of a world model displayed in a view.

5.5 Lighting

Lighting is enabled by using the *D3DMMRS_LIGHTING* status variable. If status value is TRUE, lighting is enabled and the vertex colors are generated according to the lights present in a scene. If status value is FALSE, then lighting is disabled and no light computation takes place. The default value for *D3DMMRS_LIGHTING* status is *TRUE*. A property bit, called *D3DMDEVCAPS_SPECULAR*, indicates if a device supports specular illumination management. If a device supports this feature, then, the lighting operation can be managed by adding a *D3DMRS_SPECULARENABLE* status. Also in this case, if status value is TRUE, other computations are performed on the vertices; otherwise if status value is *FALSE*, only diffuse color computations are processed and specular color is set to zero for each vertex.

D3DM supports an unlimited number of lights per scene. The only limit on the number of lights is imposed by the software drivers. Lights are identified by index values. There are many methods for managing a light in *IDirect3DMobileDevice*, such as: *IDirect3DMobileDevice::SetLight*, or *IDirect3DMobileDevice::LightEnable*. Lights are described by a *D3DMLIGHT* structure:

```
typedef struct _D3DMLIGHT {
    D3DMLIGHTTYPE Type;
    D3DMCOLORVALUE Diffuse;
    D3DMCOLORVALUE Specular;
    D3DMCOLORVALUE Ambient;
    D3DMVECTOR Position;
    D3DMVECTOR Direction;
    float Range;
    float Attenuation0;
    float Attenuation1; 89

    float Attenuation2;
} D3DMLIGHT;
```

The output of the lighting phase is diffuse and eventually the specular color of the vertices appears in the *Alpha*, *Red*, *Green*, and *Blue* (ARGB) format. If lighting computation is not enabled; the vertex colors are computed as follows:

- If a diffuse color value is already present, it is passed to the system by the *rendering pipeline*; otherwise a white color is used as the default (*0xFFFFFFFF*).
- If a specular color value is already present, it is passed to the system by the *rendering pipeline*; otherwise a black color is used as the default (0).

If lighting is enabled, the vertex colors, before being passed to a *rasterizer*, are normalized in the range $[0.0, 1.0]$ and scaled between 0 and 255. If normals (with respect to a fixed vertex) are not specified, then all dot products concerning that normals are set to zero.

As we already described for OpenGL ES API, there are four kinds of lights in D3DM: point lights, directional lights, spot lights, and ambient lights.

Point lights emit light equally in every direction from a particular point in space. This sort of light can be used, for instance, to model a typical light bulb.

Directional lights are supposed to be located an infinitely far distance away. Because of this assumption, all light rays are parallel to one another; they are all oriented in the same direction. A directional light usually models natural lighting, like sunlight, for example.

Spotlights are much like point lights, in that they emit light from a specific point in space. But rather than irradiate in all directions, they are constrained to point toward a certain direction. This kind of light shapes a cone in space.

Ambient light models lights that echo off of objects. This effect is a kind of overall lighting that illuminates everything more or less uniformly.

There are generally two kinds of attributes, that lights have: diffuse and specular, as shown in the following snippet of code.

```
protected void Lights()
 {

    dev.Lights[0].Diffuse = Color;

    dev.Lights[0].Type = LightType.Directional;

    dev.Lights[0].Direction = new Vector3(x, y, z);

    dev.Lights[0].Update();

    dev.Lights[0].Enabled = true;

 }
```

In this code we're accessing the *Lights* array of a device. Lights are usually a limited resource but they can differ depending on the systems. We can manage them individually by the *Lights* array.

First we set a diffuse color. Usually the light color is white, although other colors might be useful for different situations. Then we set the type of light. Here we are considering a directional light, but we could also have used a point light or other kinds of light by setting the appropriate parameters.

Directional lights require specifying a direction. So we set a vector for specifying direction. If, for instance, we are setting up a point light, we don't need to do the same (specify a direction), since they emit in all directions, but we need to position a light source. Spotlights, instead, would require defining both a position and a direction.

After setting up a light, we need D3DM to acquire changes (in the light attributes) by the Update method, and we turn lights on by enabling them.

After setting up the lighting, we only need to call *Lights*, for instance during the graphics context initialization.

That's another thing we must take care of; in fact, recall that lighting interacts with a surface according to its surface normal. We have already discussed surface normals, and mentioned briefly that information is stored with vertices.

To manage normals, we need to use one of the vertex formats that stores normal information, like *Vertex.PositionNormalColored*.

A constructor for a *PositionNormalColored* vertex takes the same arguments as *PositionColored*; which we have already managed, plus three more: x, y, and z components of a vertex surface normal.

Since we have decided to use a certain vertex format, we'll need to be sure we inform the device, by invoking:

```
dev.VertexFormat = Vertex.PositionNormalColored.Format;
```

If we don't set the appropriate format, the device will be expecting vertices in some other format, and this may cause errors.

5.6 Summary

This chapter explored the Microsoft Direct3D Mobile (D3DM) library, describing the architecture and rendering pipeline associated with it.

D3DM is implemented with COM (Component Object Model) interfaces and objects; COM is an object-oriented protocol used by Microsoft products that is quite powerful and extensible. Thus we introduced the basic usage of COM interfaces that are useful for developing D3DM code.

Finally, we provided snippets of code for geometric primitives and per-vertex operations, transformations, and lighting.

6

Conclusions and Prospect

This book presented an introduction to mobile 3D Graphics applications and libraries. We introduced basic graphics concepts such as: the rendering pipeline, geometric transformations, and world coordinates. These concepts have been presented by using mobile graphics libraries, thus providing a basic introduction to readers unfamiliar with graphics concepts. In the first part we discussed the limits and prospects of mobile graphics applications. We also highlighted fields and applications that can provide benefits to those employing these technologies and tools. In the second part we introduced graphics programming with the major libraries available on the market: OpenGL ES API, M3G, and Microsoft Direct 3D Mobile. By showing code samples we presented programming APIs and showed how mobile 3D graphics applications can be developed. After evaluating the presented libraries, the reader can take advantage of this information by choosing the library that best suits his application domain.

When new technologies become available, attention is paid [46]. We now look at the future possibilities for two different points of view: research and programming.

From the research point of view there is a new approach to distributed adaptive computation for improving the rendering performance of mobile devices. For instance in [47] the authors describe a Mobile Adaptive Distributed Graphics Framework (MADGRAF), which is a graphics-aware middleware architecture that enables mobile devices to run complex 3D graphics applications over wireless networks. The approach to performance problems is solved by client-server architecture, with the server performing preprocessing of complex graphics scenes and then progressively transmitting compressed graphics rendered frames tailored to the clients capabilities. This approach is related to a general approach called *Transcoding*, which is usually defined as the process of converting a media file or object from one format to another. But in this case it can be used to fit 3D graphics computations to the performance constraints of mobile devices.

Since performance issues are crucial for graphics rendering on mobile devices, being able to benchmark mobile 3D Graphics in the design cycle of that device is important. In fact, the benchmark results can be used to guide performance optimization.

In [48] the authors describe a synthetic content approach for measuring OpenGL ES 3D graphics performance of mobile devices. They developed a synthetic content tool that can create different kinds of OpenGL ES graphics content according to a large number of input parameters. The synthetic content is checked by comparing the performance of the real and synthetic contents in the same platform.

The creation of two-dimensional images of three-dimensional scenes in real time using mobile devices (or wireless devices) is becoming more and more common. Typical applications are human - computer interfaces, geographic applications, games, and more. 3D graphics on these mobile devices is a complex task due to factors like the display resolution and battery consumption. This kind of optimization is another area of research where prospective results are coming to light.

In [49] the authors present a texture compression scheme, called *iPACKMAN*. This new algorithm is an extension of the *PACKMAN* texture compression system, and while it is a bit more complex than *PACKMAN*, it is still very undemanding in terms of computational requirements but achieves very good results in the quality of rendered images.

From the programming point of view, it is hard to write about future developments since companies and groups can change their plans, but there is some news on OpenGL ES that could be of interest as an example of how libraries can split and specialize for different domains (desktop, mobile, multimedia, ...).

Most recently the Khronos organization has announced that OpenGL ES is dividing into OpenGL ES 1.x and OpenGL ES 2.x. OpenGL ES 1.x is intended for platforms like embedded and some handsets, which have less computational power. OpenGL ES 2.x is intended for platforms that can support programming [50]. OpenGL ES 2.x will not be 100 percent backward-compatible with OpenG ES 1.x. As Jon Peddie wrote in *Tech Watch* volume 5, number 6, 2005, "In general it is hoped that applications developers can deal with this through the use of OpenGL ES 2.0 shader programs, but if vendors have a product that requires both OpenGL ES 1.x and OpenGL ES 2.x they can always ship both with the product".

This is an example of the reason we chose to describe many programming approaches in this book; not only is this an introductory book, but also it is a book that will help readers in choosing the right approach, depending on their application/research area.

A

Appendix A: OpenGL®ES Code Samples

A.1 Starting with a Window

OpenGL ES provides basic functions for specifying graphics primitives, attributes, geometric transformations, viewing transformations, and many other operations. As we noted in earlier, OpenGL ES is designed to be hardware independent; therefore, many operations, such as input and output routines, are not included in this library. However, input and output routines and many additional functions are available in auxiliary libraries that have been developed for OpenGL programs.

The first step in developing an OpenGL application, and, in general, a graphics application, consists of setting up a display window.

We will present code for an example called *main.cpp*. First we will use two libraries starting with *libGLES_CM.lib*. This library is the main OpenGL ES library. The second one is *ug.lib*, and it is specific for abstracting window interface environments; this library abstracts from OS implementation of windowing systems, thus behaving like the OpenGL utility toolkit (GLUT) for standard OpenGL®, and freeing developers from supporting a specific OS. In fact, in addition to OpenGL ES basic library, there are a number of associated libraries for handling special operations. The *ug.lib* provides a library of functions for interacting with any screen-windowing system.

It's possible to link these libraries using our specific integrated development environment (IDE) or developing environment, but we will use a standard technique based on the *Cpragma* statement.

To link a library we use the following syntax:
`♯pragma comment(lib, "LIBRARY_NAME")`

```
#pragma comment(lib, "libGLES\_CM.lib")
#pragma comment(lib,"ug.lib")
```

In our graphics programs, we need to include a header file for the OpenGL ES core library. For most applications we also need *ug.lib* for the windowing system. Since we are using the *ug.lib* for abstracting interfaces from the corresponding OS, we only need to include a *GLES/gl.h* file as it includes also *GLES/egl.h* and all the needed OpenGL ES functions.

```
#include "ug.h"
```

In future examples all initialization code is included in an *init* function, which in this first example is empty.

```
void init()
{
}
```

To create a graphics content using OpenGL ES API, we first need to set up a *display window* on our video screen. This is simply a rectangular area of the screen in which our pictures will be displayed. We cannot create a display window directly with basic OpenGL functions, since this library contains only device-independent graphics functions, and window-management operations depend on the supported OS.

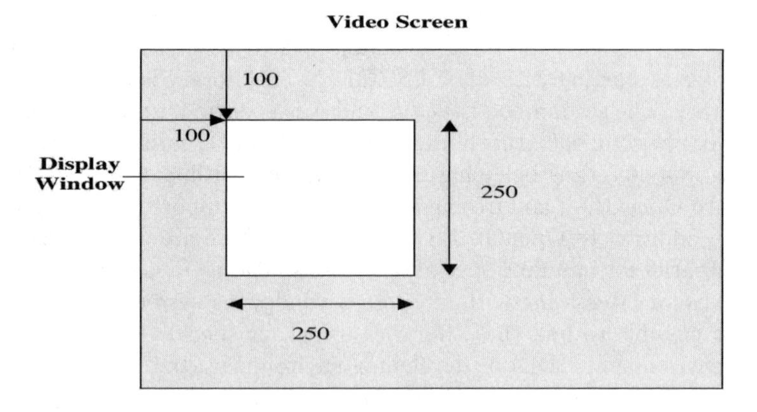

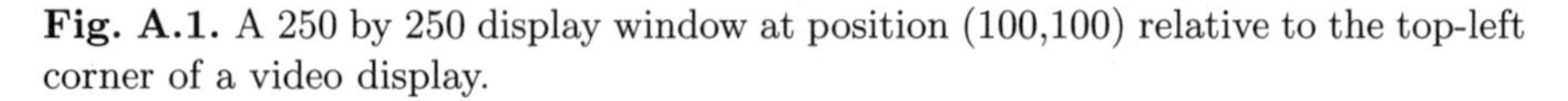

Fig. A.1. A 250 by 250 display window at position (100,100) relative to the top-left corner of a video display.

The OpenGL paradigm displays graphics on the screen by using frames. For each single frame it is necessary to develop what will be presented on the screen. A display function is executed for each frame; thus all graphics code will be inserted in this function. It requires an input parameter of the *UGWindow* type, which represents the displayed window.

```
void display(UGWindow window) {

}
```

Now let's focus on the main routine. The first variable to be considered is *UGCTx*. *UGCTx* is a handle that manages a default graphic engine (*UG*). The **ugInit** function initializes a graphics engine and returns a handle assigned to the *UGCTx* variable.

```
int main() {


    UGCTx HandleEngUg = ugInit();
```

The next step consists of creating a window using the **ugCreateWindow** function. As already shown by the display function, a window variable *UGWindow* is used to store a handle to an OpenGL ES window. The function parameters are:

UGCTx **ug**: an handle.
const *char* ***config**: for managing options. This action enables some buffers that will be described later; for now we can leave it as an empty string.
const *char* ***title**: this specifies the text being displayed as the window title on top of a window.
int **width** & *int* **height**: the width and height of a window.
int **x**& *int* **y**: the left top screen corner coordinates where a displayed window will be positioned.

```
UGWindow window = ugCreateWindow(HandleEngUg, "",

"Create Window", 250, 250, 100, 100);
```

Now we call the already declared **init** function.

```
init();
```

Then we send to the OpenGL window objects to be displayed. We can do this by using a display function, *ugDisplayFunc*, which takes as input a window and one display handle.

```
ugdisplayFunc(window, display);
```

To prevent a program from immediately stopping, we need a sort of a loop. This is what we actually call the *Main Loop*. This loop continue to iterate, managing program messages and/or events. We can call the **ugMainLoop** function only with a parameter, the window handle.

```
ugMainLoop(ug);

return 0;

}
```

We showed the basic steps for creating a simple window with OpenGL ES API. Running this program, it seems that nothing has happened, and in fact we haven't yet specified what geometric primitives to display on the screen. You'll notice that pushing the *OK* button, in the top right corner, the program will continue to run. The next paragraph shows how to manage interfaces using the keyboard and mouse, and thus also how to quit a main program.

A.2 Basic Interaction

Many programs require inputting data from the keyboard or mouse; thus we need a way to associate an action with the key pressed.

The first step in managing keyboard inputs is to define a function that has some parameters in input. We will define a basic function accepting four parameters:

1. Current window linked by a *UGwindow* variable.
2. Key pressed, stored in an integer variable.
3. X coordinate of the mouse pointer when a key has been pressed.
4. Y coordinate of the mouse pointer when a key has been pressed.

```
void keyboard(UGwindow uwin, int key, int x, int y) {
```

Then we check which key has been pressed.

```
switch(key) {
```

We can check a *key* variable with each kind of character. In our example the *'q'* character means quit and exit from the program.

```
case 'q': exit(0); break; }
}
```

Standard characters can be test enclosed in quotation marks, while other characters are identified by special key codes as shown in Table A.1.

Identifier	Description
UG_KEY_F1 - UG_KEY_F2	function key from F1 to F12
UG_KEY_LEFT	left arrow
UG_KEY_RIGHT	right arrow
UG_KEY_UP	up arrow
UG_KEY_DOWN	down arrow
UG_KEY_PAGE_UP	page up
UG_KEY_PAGE_DOWN	page down
UG_KEY_HOME	home key
UG_KEY_END	end key
UG_KEY_INSERT	Ins key

Table A.1. Special key codes.

We then link the keyboard input function to the main window in order to manage the key pressing events in that specific window. We call the *ugKeyboardFunc* function at the same point of the program where we call *ugDispalyFunc. ugKeyboardFunc* takes a window manager and keyboard function as parameters and links them.

```
ugKeyboardFunc(uwin, keyboard);
```

This is a simple example but shows how to manage keyboard interactions, and by executing this code fragments in the main code, explained above, you can exit from the displayed window pressing the "q" key.

A.3 Geometric primitives and Per-Vertex Operations

In this example we see how to display a shape on the screen. Geometric primitives, like squares or triangles, are implemented by specifying vertices of geometric shapes. Vertices are points in a three-dimensional space, and thus are composed of three coordinates: x, y, and z. After specifying the vertices, it is necessary to specify the types of geometric primitive, listed in the Chapter 3, section 3.3.3, table 3.1.

To display a triangle, one must specify coordinates of vertices for that triangle, as follows in the code below.

```
GLfloat Triangle[] = {
    0.25f, 0.65f, 0.0f,
    0.35f, 0.35f, 0.0f,
    0.85f, 0.85f, 0.0f,
    };
```

We then define an *init* function for initializing the code. A graphic window is initialized by declaring its background color in red, green, blu, and alpha transparency (RGBA) by **glClearColor** function.

```
void init() {

    glClearColor(1.0f,0.4f,0.4f,0.0f);
```

Recall from Chapter 3, section 3.3.3 that in OpenGL ES transformations are are managed by the *GL_MODELVIEW* matrix, while views are managed by *GL_PROJECTION*. We use the projection matrix in this example.

There are many different specialized arrays in OpenGL ES that are, by default, disabled. We enable only arrays we need; in this case, we enable only color and vertex arrays by using the **glEnableClientState**.

```
    glClearColor (1.0f, 0.4f, 0.4f, 0.0f);
    glEnableClientState(GL_COLOR_ARRAY);
    glEnableClientState(GL_VERTEX_ARRAY);
    glColorPointer(4, GL_FLOAT, 0, Colors);
    glShadeModel(GL_FLAT);
    glMatrixMode(GL_PROJECTION);
```

We initialize (clear) the projection matrix by using an identity matrix [1].

```
glLoadIdentity();
```

[1] The identity matrix of size n is the n-by-n square matrix with ones on the main diagonal and zeros elsewhere. In particular, the identity matrix serves as the unit of all n-by-n matrices.

To compute orthogonal projections we use the **glOrthof** function. It requires six parameters for specifying clipping planes (left, right, bottom, top, near, and far), as shown in Chapter 3, Figure 3.3. These parameters span the area (of the scene) to be visualized in our display window.

```
glOrthof(0.0f, 1.0f, 0.0f, 1.0f, -1.0f, 1.0f);
```

To display the triangle we define a view port (i.e., a window on the screen) by setting top-left and bottom-right corner coordinates. Finally, by the **glVertexPointer** we load in the vertex array, the triangle shape stored in the *Triangle* matrix.

```
glViewport(0, 0, 250, 250);

glVertexPointer(3, GL_FLOAT, 0, square);

}
```

For now we have set the view (orthographic) and the shape (a triangle). To display a shape, we clear the screen by the **glClear** function.

```
void display(UGWindow uwin)

{

glClear(GL_COLOR_BUFFER_BIT);
```

To draw a geometric primitive, OpenGL ES use the current vertex array; it is used, as a parameter, by the **glDrawArrays** function; this function requires, also, to specify how many vertices are need to draw the shape and what type of geometric primitive to use (*GL_TRIANGLE_STRIP* in our example).

```
glDrawArrays(GL_TRIANGLE_STRIP, 0, 3);
```

We finish by sending data from memory buffers (color and vertex arrays) to the screen.

```
glFlush();

ugSwapBuffers(uwin);

}
```

Fig. A.2. An orthogonal view of a square made by two triangles stripped together.

The displayed image will look like Figure A.2.

We now describe another example related to per-vertex operations; it consists of managing primitive transformations (rotation, translation, and scaling).

To show rotation (as an example of transformation) we will use a simple animation, and that is also interesting as it describes the OpenGL ES capabilities at rendering time.

First we introduce two variables that hold the status of current rotation with respect to the x and y axis.

```
float xrot = 0.0f;

float yrot = 0.0f;
```

We then introduce an array containing triangle vertices, as already seen in a previous example. We are defining primitives as centered in axis origin, and in fact our transformation manages our model, a cube in this case, with respect to the axis origin, as described below.

```
float val = 0,3f;

GLfloat Cube[] = {
    // Front
    -val, -val,  val,
     val, -val,  val,
    -val,  val,  val,
     val,  val,  val,
    // Rear
    -val, -val, -val,
    -val,  val, -val,
     val, -val, -val,
     val,  val, -val,
    // Left side
    -val, -val,  val,
    -val,  val,  val,
    -val, -val, -val,
    -val,  val, -val,
    // Right side
     val, -val, -val,
     val,  val, -val,
     val, -val,  val,
     val,  val,  val,
    // Up
    -val,  val,  val,
     val,  val,  val,
    -val,  val, -val,
     val,  val, -val,
    // Down
    -val, -val,  val,
    -val, -val, -val,
     val, -val,  val,
     val, -val, -val,
};
```

We then define a color array. Colors are specified by a group of four coordinates in RGBA space.

```
GLfloat Colors[] = {
    1.0f, 0.0f, 0.0f, 1.0f,
    0.0f, 1.0f, 0.0f, 1.0f,
    0.0f, 0.0f, 1.0f, 1.0f
};
```

The initialization function simply sets a background color.

```
void init()

{

glClearColor(0.87f,0.87f,0.87f,0.0f);

}
```

The display function is define as usual.

```
void display(UGWindow uwin)

{

glClear(GL_COLOR_BUFFER_BIT);
```

We now draw the cube on the screen. It will include a smooth color. There are two kinds of shading implemented by the **glShadeModel** function: GL_FLAT and GL_SMOOTH. GL_SMOOTH is default shade model. GL_FLAT makes a shape single color. GL_SMOOTH enables smooth shading, that is, vertices and primitive colors are computed by interpolating single values of vertex colors.

```
glShadeModel(GL_SMOOTH);

glVertexPointer(3, GL_FLOAT, 0, cube);

glColorPointer(4, GL_FLOAT, 0, colors);

glEnableClientState(GL_VERTEX_ARRAY);

glEnableClientState(GL_COLOR_ARRAY);
```

Transformations are executed by the functions **glTranslatef**, **glScalef**, and **glRotatef**. The f in each function name indicates that these functions accept only floating point parameters as input. Usually an x indicates a GLfixed type for input parameters, while v is used for arrays.

After drawing the cube, we don't want other shapes to be influenced by the next transformations, since that will change the geometric coordinate reference. the **glPushMatrix** and **glPopMatrix** functions are used for storing the actual state of the reference system in a stack. We insert our transformation code between these two functions, thus being sure that the *model-view* matrix will be reset to its original state (reference system) after execution of that code segment.

```
glPushMatrix();
```

We now apply all three transformations. Each transformation will change the *model-view* matrix, thus influencing the successive transformations. Recalling that geometric transformations are carried out by matrix multiplication, a product in matrix algebra changes with the order of factors, in our example, translating a shape (geometric primitives) to right side of screen and rotating. While rotating and then translating, we will see the shape move diagonally with respect to the origin. To test this, try to change the order of the following transformations.

glTranslatef takes three parameters as input, thus indicating how to move along all three axis. Our first transformation consists of moving a cube 0.25 units right and 0.5 to the top.

```
glTranslatef(0.5f, 0.5f, 0.0f);
```

Now we scale our object, recalling that is centered in the origin. The scaling functions takes three values for each vertex and multiplies them for its input parameters. If a shape was placed in the bottom left of a window, this operation will have scaled the objects but the origin would still be centered in the bottom left angle.

Thus we reduce our cubic shape by three fourths. It is possible to have a different scaling coefficient for each axis.

```
glScalef(0.75f, 0.75f, 0.75f);
```

Finally, rotation takes place. The first parameter specifies the rotation angle. The other three are used for indicating with respect to which axis an object is rotated. A value of 1.0 usually indicates an axis.

```
glRotatef(xrot, 1.0f, 0.0f, 0.0f);
```

We will now draw the cube. It will appear in three-fourth on the left half of the screen.

```
// Back and Rear
   glColor4f(1.0f, 0.0f, 0.0f, 1.0f);
   glNormal3f(0.0f, 0.0f, 1.0f);
   glDrawArrays(GL_TRIANGLE_STRIP, 0, 4);
   glNormal3f(0.0f, 0.0f, -1.0f);
   glDrawArrays(GL_TRIANGLE_STRIP, 4, 4);

   // Left and Right sides
   glColor4f(0.0f, 1.0f, 0.0f, 1.0f);
   glNormal3f(-1.0f, 0.0f, 0.0f);
   glDrawArrays(GL_TRIANGLE_STRIP, 8, 4);
   glNormal3f(1.0f, 0.0f, 0.0f);
   glDrawArrays(GL_TRIANGLE_STRIP, 12, 4);

   // Up and Down
   glColor4f(0.0f, 0.0f, 1.0f, 1.0f);
   glNormal3f(0.0f, 1.0f, 0.0f);
   glDrawArrays(GL_TRIANGLE_STRIP, 16, 4);
   glNormal3f(0.0f, -1.0f, 0.0f);
   glDrawArrays(GL_TRIANGLE_STRIP, 20, 4);
```

We now restore the original reference system by taking out from the stack the original matrices.

```
glPopMatrix();
```

We disable the color array, because it won't be used for the rest of this example.

```
glDisableClientState(GL_COLOR_ARRAY);
```

And we close as usual.

```
glFlush();

ugSwapBuffers(uwin);

}
```

To include an animation mechanism, we need a function called **idle**. This function is called in the main loop when there are no other commands to execute. It takes as parameter a window handler. This function increments the rotation angle on shapes with respect to the x and y axes. Moreover, we set the screen to be refreshed after changing the values, and this is obtained by the **ugPostRedisplay** function.

```
void idle(UGWindow uwin) { xrot += 1.0f; yrot += 1.0f;
ugPostRedisplay(uwin); }
```

The last thing to do is to specify to the **UG** engine which kind of **idle** function to use via **ugIdleFunction**. This function takes two parameters: a **UG** engine handle and an **idle** function. It is placed where **ugMainLoop** is called.

```
ugIdleFunc(ug, idle);
```

The displayed image will look like Figure A.3.

A.4 Lighting

We start by defining two color arrays, one for ambient light and another for diffuse light. The last one represents a color for the lighting source.

```
float lightAmbient[] = { 0.5f, 0.5f, 0.5f, 1.0f };

float lightDiffuse[] = { 0.5f, 0.5f, 0.5f, 1.0f };
```

Now an array for specifying materialproperties is needed, one for ambient and another for diffuse light. Basically we multiply the lighting values by the material values in order to obtain a final reflected color. Each value represent a quantity used for reflecting a particular color.

```
float matAmbient[] = { 1.0f, 0.0f, 0.0f, 1.0f };

float matDiffuse[] = { 1.0f, 0.0f, 0.0f, 1.0f };

void init() {
```

First let's activate lighting by using the GL_LIGHTING parameter as input for the **glEnable** function.

Fig. A.3. The rotated cube as an example of geometric primitives transformations.

```
glEnable(GL_LIGHTING);

glEnable(GL_COLOR_MATERIAL);
```

OpenGL ES allows the use of eight different lights at the same time. To enable one of these lights, a GL_LIGHTx parameter has to be passed to the **glEnable** function as input, with $x = 0 \ldots 7$.

```
glEnable(GL_LIGHT0);
```

To define material properties, we use the **glMaterialfv** and **glMaterialf** functions. **glMaterialfv** is used for multiple valued parameters, while **glMaterialf** is used when there is a single valued parameter, as shown later in this example.

The first parameter defines which polygon face needs to be updated by lighting information (for example GL_FRONT). The second parameter is used to specify the type of lighting attributes and thus could be GL_AMBIENT, GL_DIFFUSE, GL_SPECULAR, GL_EMISSION, or GL_AMBIENT_AND_DIFFUSE.

The last parameter is an array or single value depending on the function used (**glMaterialfv** or **glMaterialf**).

The next two lines set the material properties.

```
glMaterialfv(GL_FRONT_AND_BACK, GL_AMBIENT, matAmbient);

glMaterialfv(GL_FRONT_AND_BACK, GL_DIFFUSE, matDiffuse);
```

Also lighting properties have to be set, and this can be done by using the **glLightfv** and **glLightf** functions, which work in the same manner as material functions.

```
glLightfv(GL_LIGHT0, GL_AMBIENT, lightAmbient);

glLightfv(GL_LIGHT0, GL_DIFFUSE, lightDiffuse);
```

The remaining code for *init* functions is shown below.

```
glEnable(GL_DEPTH_TEST);

glDepthFunc(GL_LEQUAL);

glClearDepthf(1.0f);

glClearColor(0.87f, 0.87f, 0.87f, 0.0f);

glVertexPointer(3, GL_FLOAT, 0, box);

glEnableClientState(GL_VERTEX_ARRAY);

glEnable(GL_CULL_FACE);

glShadeModel(GL_SMOOTH);

}
```

The display function is the same as in the other examples.

```
void display(UGWindow uwin) {

glClear(GL_COLOR_BUFFER_BIT | GL_DEPTH_BUFFER_BIT);

glLoadIdentity();

    ugluLookAtf(
           0.0f, 0.0f, 3.0f,
           0.0f, 0.0f, 0.0f,
           0.0f, 1.0f, 0.0f);

glRotatef(xrot, 1.0f, 0.0f, 0.0f);

glRotatef(yrot, 0.0f, 0.0f, 1.0f);
```

We already defined the normals, and they must be perpendicular to the surfaces. Thus a surface in front of a light must have a *(0,0,1)* normal vector, while a back surface has *(0,0,-1)*. The vector length is one; thus both are normalized vectors.

Normals are defined by the *glNormal3F* function before drawing the related primitives, and this function takes as input three parameters that identify the normalized vectors.

```
//FRONT AND BACK

glColor4f(1.0f, 0.0f,0.0f,1.0f);
glNormal3f(0.0f, 0.0f, 1.0f);
glDrawArrays(GL_TRIANGLE_STRIP, 0, 4);
glNormal3f(0.0f, 0.0f, -1.0f);
glDrawArrays(GL_TRIANGLE_STRIP, 4, 4);
```

The same thing is done for the bottom and side surfaces. Like color and vertex arrays, there is also a normal array. It can be initialized by the **glNormalPointer** function, which works just like **glVertexPointer**.

To enable this array, the GL_NORMAL_ARRAY flag must be passed to **glEnableClientState** as input.

```
//LEFT AND RIGHT
```

```
    glColor4f(0.0f, 1.0f,0.0f,1.0f);
    glNormal3f(-1.0f, 0.0f, 0.0f);
    glDrawArrays(GL_TRIANGLE_STRIP, 8, 4);
    glNormal3f(1.0f, 0.0f, 0.0f);
    glDrawArrays(GL_TRIANGLE_STRIP, 12, 4);

    //TOP AND BOTTOM

    glColor4f(0.0f, 0.0f, 1.0f,1.0f);
    glNormal3f(0.0f, 1.0f, 0.0f);
    glDrawArrays(GL_TRIANGLE_STRIP, 16, 4);
    glNormal3f(0.0f, -1.0f, 0.0f);
    glDrawArrays(GL_TRIANGLE_STRIP, 20, 4);
    glFlush();
    ugSwapBuffers(uwin);

}
```

Figure A.4 shows an example of lighting.

In the first example we enriched our scene by including lighting. The included light didn't have a particular direction, though. We will see now how to use directional lights; this will allow us to manage diffuse and specular illumination.

To better detect a visual effect of specular lighting, we set in the center of our scene a red ball with an intense light pointing at it.

First let's create arrays for setting the light properties and add a specular array for a specular effect.

```
float lightAmbient[] = { 0.5f, 0.5f, 0.5f, 1.0f };

float lightDiffuse[] = { 0.5f, 0.5f, 0.5f, 1.0f };

float lightSpecular[] = { 0.5f, 0.5f, 0.5f, 1.0f };
```

We now create a specular array also for a material. Let's set it so that a material will reflect all lights that hit it.

```
float matAmbient[] = { 1.0f, 0.0f, 0.0f, 0.0f };

float matDiffuse[] = { 1.0f, 0.0f, 0.0f, 0.0f };

float matSpecular[] = { 1.0f, 0.0f, 0.0f, 0.0f };
```

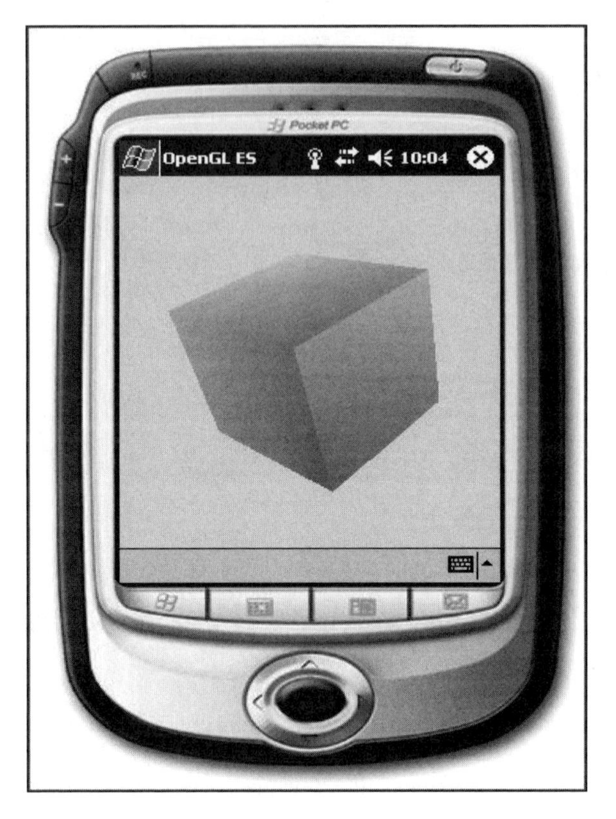

Fig. A.4. The scene with the color tracking enabled.

Since we're dealing with a directional light, we must set the light position and direction. We create two arrays for specifying these two properties. We choose a sphere as the geometric model.

```
float lightPosition[] = { 200.0f, 0.0f, 0.0f, 1.0f };

float lightDirection[] = { -200.0f, 0.0f, 0.0f, 1.0f };
```

We now enable lighting and the first light.

```
void init()

{
    glEnable(GL_LIGHTING);

    glEnable(GL_LIGHT0);
```

We set the material properties and a specular value.

```
glMaterialfv(GL_FRONT, GL_AMBIENT, matAmbient);

glMaterialfv(GL_FRONT, GL_DIFFUSE, matDiffuse);

glMaterialfv(GL_FRONT, GL_SPECULAR, matSpecular);
```

We then set a new material property by using the **glMaterialf** function. The shininess value for material is usually in the $[0, 128]$ range. This value specifies how much specular light will be polarized. The greater the value, the more the light will be polarized.

```
glMaterialf(GL_FRONT, GL_SHININESS, 20.0f);
```

The next step consists of setting the light properties.

```
glLightfv(GL_LIGHT0, GL_AMBIENT, lightAmbient);

glLightfv(GL_LIGHT0, GL_DIFFUSE, lightDiffuse);

glLightfv(GL_LIGHT0, GL_SPECULAR, lightSpecular);
```

To set a position and the direction of the light, the *GL_POSITION* and *GL_SPOT_DIRECTION* flags must be set and passed as input to a **glLightfv** function.

```
glLightfv(GL_LIGHT0, GL_POSITION, lightPosition);

glLightfv(GL_LIGHT0, GL_SPOT_DIRECTION, lightDirection);
```

Another useful flag is *GL_SPOT_CUTOFF*. It specifies a light cone size. We can imagine an effect that is like an electric torch cone pointing to a wall. For instance, a value of 1.2 creates a cone with an angle of 2.4 degrees. A value of 180 will spread light in every direction.

```
glLightf(GL_LIGHT0, GL_SPOT_CUTOFF, 10f);
```

Finally, there three more flags that can be used:

- *GL_CONSTANT_ATTENUATION*,

- *GL_LINEAR_ATTENUATION*,
- *GL_QUADRATIC_ATTENUATION*.

They can be used to manage light reduction.

Light reduction is a measure of how much light intensity is reduced by moving far from a light source. In the torch example, its effect is light reduction when moving far from the torch itself. We must consider that setting these properties could end in decreasing the software performance since they require many calculations and thus we won't use them in our example.

```
    glEnable(GL_DEPTH_TEST);

    glDepthFunc(GL_LEQUAL);

    glClearDepthf(1.0f);

    glClearColor(0.87f, 0.87f, 0.87f, 1.0f);

    glEnable(GL_CULL_FACE);

    glShadeModel(GL_SMOOTH);

}
```

We now create our ball (sphere) by using the **ugSolidSpheref** function. At this point one may ask, where are the normal arrays set? The answer is that the **UG** library automatically computes the normal directions and values.

```
void display(UGWindow uwin)

{

    glClear(GL_COLOR_BUFFER_BIT | GL_DEPTH_BUFFER_BIT);

    glLoadIdentity();

    ugluLookAtf(
        0.0f, 0.0f, 4.0f,
        0.0f, 0.0f, 0.0f,
        0.0f, 1.0f, 0.0f);

    glRotatef(xrot, 1.0f, 0.0f, 0.0f);
```

```
    glRotatef(yrot, 0.0f, 1.0f, 0.0f);

    ugSolidSpheref(1.0f, 20, 20);

    glFlush();

    ugSwapBuffers(uwin);

}
```

The final scene rendering displays a red sphere with a specular reflection on the top right side of the sphere (Figure A.5).

Fig. A.5. A red sphere with a specular reflection on top right side of the sphere.

References

1. Pulli, K., The rise of mobile graphics. Information Quarterly–IQ Magazine (www.iqmagazineonline.com), (1), 2004.
2. Azuma, R., A survey of augmented reality. In: Presence: Teleoperators and Virtual Environments, Cambridge. MA: MIT Press, 6(4):355–385, (1997).
3. Pham, B., Wong, O., Handheld devices for applications using dynamic multimedia data. In Proceedings of the 2nd international Conference on Computer Graphics and interactive Techniques in Australasia and South East, Asia, Singapore, June 15-18, 2004, S. N. Spencer, Ed. GRAPHITE '04. New York, ACM Press, NY, 123–130, 2004.
4. Yao, B., Fuchs, W. K., Recovery proxy for wireless applications. Proceedings of ISSRE2001, 112–119, Nov. 2001.
5. Frohlich, B., Plate, J., The cubic mouse: a new device for three-dimensional input. In: Turner, T., Szwillus, G., Czerwinski, M., Peterno, F., Pemberton, S. (eds.): Proceedings of the ACM CHI 2000 Human Factors in Computing Systems Conference, April 1–6, 2000, The Hague, The Netherlands, pp. 526–531.
6. Rekimoto, J., Sciammarella, E., ToolStone: effective use of the physical manipulation vocabularies of input devices. 109–117, UIST 2000 Symposium on User Interface Software and Technology.
7. Foley, J.D., Van Dam, A., Feiner, S.K., Hughes, J.F., Computer Graphics: Principles and Practice in C, 2nd Ed. New York: Addison-Wesley Professional, 1995.
8. Baker, H., Computer Graphics with OpenGL, 3rd Ed. Upper Saddle River, nJ. Pearson Prentice-Hall, 2004.
9. Molofee, J., OpenGL Tutorial, by Neon Helium Productions, nehe.gamedev.net.
10. Sutherland, I.E., Sproull, R.F., Schumacker, R.A., A characterization of ten hidden-surface algorithms. ACM Comput. Surv. 6(1):1–55, 1974.
11. Baker, S., Basic OpenGL Lighting. www.sjbaker.org/steve/omniv/opengl_lighting.html.
12. Goodwin, D., Small and beautiful, feature, March 2004, Khronos group, www.khronos.org/news/articles.
13. Introduction to the Mobile 3-D Graphics API For J2ME, from Forum Nokia, www.forum.nokia.com.
14. Vainio, T., Kotala, O., Rakkolainen, I., Kupila, H., Towards scalable user interfaces in 3D city information systems. In Proceedings of the 4th International Symposium on Mobile Human-Computer Interaction, September 18–20,

2002, F. Paterno', ed. Lecture Notes in Computer Science, vol. 2411. London: Springer–Verlag, 354–358, 2002.

15. Burigat S., Chittaro L., Location-aware visualization of VRML models in GPS-based mobile guides, Proceedings of Web3D 2005: 10th International Conference on 3D Web Technology. Ne York: ACM Press, 57–64, 2005.
16. http://earth.google.com/.
17. Milgram, P., Kishino. F., A taxonomy of mixed reality visual displays. IEICE Transactions on Information Systems, Vol E77-D, No. 12, December 1994.
18. Wagner D., Schmalstieg D., First steps towards handheld augmented reality. In Proceedings of the 7th International Conference on Wearable Computers, White Plains, NY, Oct. 21–23, 2003.
19. Wagner D., Pintaric T., Ledermann F., Schmalstieg D., Towards Massively multi-user augmented reality on handheld devices. Lecture Notes in Computer Science, Volume 3468, 208–219, 2005.
20. International Game Developers Association. http://www.igda.org.
21. Hinter Wars: Multiplayer Mobile/Online Games. http://www.hinterwars.com.
22. Jakob Nielsen's Alertbox, September 16, 2001: Mobile Devices Will Soon Be Useful. http://www.useit.com/.
23. Bjork, S., Bretan, I., Danielsson, R., Karlgren, J., Franzen, K., WEST: A Web browser for small terminals. CHI Letters, (1), 187–196, 1999.
24. Boguraev, B., Bellamy, R., Swart, C., Point-of-view: custom information delivery via hand-held devices. Proceedings of the 34th Hawaii International Conference on System Science IEEE, 1–10, 2001.
25. Li, C.-S., Mohan, R., Smith, J. R., Multimedia content description in the InfoPyramid. Proceedings of the 1998 IEEE International Conference on Acoustics, Speech, and Signal Processing ICASSP, 1998.
26. Curtis, K., Draper, O., Multimedia content management-provision of validation and personalisation services. Proceedings of IEEE International Conference on Multimedia Computing and Systems, 1999.
27. De Rosa, F., Malizia, A., Mecella, M., Disconnection prediction in mobile ad hoc networks for supporting cooperative work. IEEE Pervasive Computing 4(3), 62–70, 2005.
28. techpubs.sgi.com.
29. Moller, T. A., Strom, J. 2003. Graphics for the masses: a hardware rasterization architecture for mobile phones. ACM Trans. Graph. 22(3), 801–808, 2003.
30. HRAA: high-resolution antialiasing through multisampling. Technical brief, NVIDIA Corp., 2001.
31. Strom, J., Moller., T. A., PACKMAN: texture compression for mobile phones. Technical sketch at SIGGRAPH, 2004.
32. Antochi, I., Juurlink, B.H.H., Vassiliadis, S., Liuha, P., Memory bandwidth requirements of tile-based rendering. In Proceedings of the Third and Fourth International Workshops SAMOS 2003 and SAMOS 2004 (LNCS 3133), pp. 323-332, Samos, Greece, 2004.
33. Antochi, I., Juurlink, B.H.H., Vassiliadis, S., Liuha, P., Efficient tile-Aware bounding-box overlap test for tile based rendering. In Proceedings 2004 International Symposium on System-on-Chip, pp. 165–168, Tampere, Finland, 2004.
34. Antochi, I., Juurlink, B.H.H., Vassiliadis, S., Liuha, P., Scene management models and overlap tests for tile-based rendering.In Proceedings of the EUROMICRO Symposium on Digital System Design 2004 (DSD 2004), pp. 424–431, Rennes, France, 2004.

35. www.falloutsoftware.com.
36. www.gamedev.net.
37. en.wikipedia.org/wiki/Antialiasing.
38. Crow, F. C., The aliasing problem in computer-generated shaded images, Communications of the ACM, vol. 20(11), November 1977, pp. 799-805.
39. www.opengl.org.
40. Mobile 3D Graphics for J2ME (JSR-184), www.jcp.org/en/jsr/detail?id=184.
41. CLDC 1.1 Connected Limited Device Configuration 1.1 (JSR-139), Java Community Process, 2003, www.jcp.org/en/jsr/detail?id=139.
42. Mobile Information Device Profile (JSR-37), Java Community Process, 2000, www.jcp.org/en/jsr/detail?id=37.
43. Williams, C., Burge, M., MIDP 2.0 changing the face of J2ME gaming, Proceedings of the 42nd annual Southeast regional conference, pp. 31–47, Huntsville, AL: ACM Press, 2004.
44. fivedots.coe.psu.ac.th/~ad/jg/objm3g.
45. Microsoft Corporation, Microsoft Developer Network, msdn.microsoft.com/default.asp.
46. Olsen, K.A., *Formalizing Internet, Web and eBusiness Applications for the Real World.* Lahnam MD: Scarecrow Press, 2005.
47. Agu, E., Banerjee, K., Nilekar, S., Rekutin, O., Kramer, D., A middleware architecture for Mobile 3D Graphics. In Proceedings of the Third international Workshop on Mobile Distributed Computing (Mdc) (Icdcsw'05), Volume 6 (June 6–10, 2005). ICDCSW, IEEE Computer Society, Washington, DC, 617–623.
48. Kangas, K. J. , Qvist M., Pulli M., Synthetic content approach for benchmarking mobile 3D graphics. SIGRAD 2005, Lund, Sweden, November 2005.
49. Strom, J., Moller, T. A., iPACKMAN: high-quality, low-complexity texture compression for mobile phones. Graphics Hardware, 63–70, 2005.
50. Jon Peddie's Tech Watch, March 28, 5(6), 2005. *www.jonpeddie.com/TechWatch.shtml.*

Index

2D, 11, 23, 27, 29, 30, 36, 38, 55, 60, 95, 97, 106
3D, 7, 8, 11, 13, 15, 16, 20–27, 29–31, 33, 34, 36–38, 41, 43, 44, 55, 64, 69, 76, 78, 86–88, 92, 95, 101, 103, 113, 115, 122, 128, 129

algorithm, 20, 43, 46, 48, 51, 78, 108, 129
alpha, 62
animated, 97–99
animation, 11, 22, 93, 98, 99, 137, 141
API, 5, 7, 8, 11, 12, 22–27, 38, 43, 55, 56, 86, 87, 89, 100–102, 104, 113–117
API., 7
APIs, 128
AR, 31, 33, 34, 38
architecture, 42, 43, 46, 49–52, 115, 127, 128
array, 11, 14, 61, 63–65, 69–73, 76, 96, 97, 119, 120, 127, 137, 138, 141, 142, 144, 145, 147
arrays, 63, 64, 66, 69, 71–73, 83, 139, 142, 145, 147, 149
Augmented, 31–33
augmented, 9, 29, 31–33, 38, 52
axes, 64, 65, 68
axis, 32, 95, 104, 106, 111, 123, 124, 137, 138, 140

background, 62, 65, 78, 91, 92, 104, 120, 135, 139
Blending, 79
blending, 27, 79–82
buffer, 13–15, 43, 46, 49–51, 69, 76, 79, 83, 116–118, 120–122
build, 12, 27, 38, 92, 93, 106

C, 22, 56, 57, 130
C++, 7, 11
camera, 11, 12, 27, 33, 38, 40, 46, 78, 94, 103, 104, 121, 124
channel, 76, 105
client-state, 83
collision, 107
collisions, 34, 100, 106, 107
colors, 3, 6, 11, 15, 23, 27, 48, 66, 68, 76, 78, 79, 82, 90, 96, 117, 125–127, 139
COM, 115, 116, 127
component, 50, 68, 77, 81
components, 48, 68, 76, 78, 81, 86, 89, 92, 94, 96–98, 115, 127
compression, 41, 46, 48, 129
coordinate, 12, 13, 27, 59, 78, 79, 94, 99, 123, 124, 133, 140
coordinates, 12–14, 20, 47, 60, 63, 65, 66, 68, 76, 78, 79, 82, 94, 96, 97, 117, 119–124, 128, 132, 134, 138
CPU, 25, 39, 42, 43, 47, 49, 55, 96
cube, 122, 138–140

D3DM, 115–121, 123–127
depth, 16–19, 43, 47, 61, 76, 78
design, 4, 5, 7, 9, 11, 23, 35–39, 41–43, 55, 87, 88, 92, 129

desktop, 3–7, 25, 31, 34–37, 42, 43, 56, 87, 115
develop, 7, 22, 24, 25, 38, 58, 60, 131
development, 4, 6–8, 11, 23, 24, 33–36, 38, 86, 88, 90, 113, 117
device, 3–5, 7–9, 12, 13, 25, 26, 36–43, 52, 88, 102, 103, 117, 119, 120, 125, 127, 129
devices, 3–10, 12, 13, 23–29, 33–44, 48, 49, 52, 55, 60, 86, 91, 113, 115, 116, 128, 129
diffuse, 20, 56, 68, 69, 72, 117, 125–127, 142, 147
digital, 3, 39
dimensional, 16, 92
direct, 9, 13, 43, 89
Direct3D, 116, 127
display, 7, 11–13, 27, 30–34, 56–58, 60, 64, 66, 69, 70, 75, 76, 83, 90, 95, 116, 117, 120, 121, 124, 129–134, 139
displays, 4, 6, 7, 30–34, 41, 58, 150
draw, 11, 39, 46, 61, 63, 64, 68, 78, 91, 121, 139, 140
drawing, 11, 12, 66, 71, 77, 91, 92, 103, 107, 116–118, 121, 140, 145

engine, 25, 38, 58, 132, 142
entertainment, 34–37
ES, 23–27, 129

face, 18–20, 40, 78
faces, 16, 18–20
features, 4–7, 23, 25, 27, 28, 36–38, 43, 44, 72, 90, 116
fixed, 40, 43–46, 59, 82, 83, 94, 99, 100, 121, 126
float, 43, 60, 83, 101, 109
floating, 24, 42–44, 55, 63, 66, 121, 139
fragment, 15, 50, 79, 81, 82
fragments, 15, 50, 59, 76, 82, 134
frame, 9, 12, 14, 15, 19, 27, 36, 39, 41, 43, 46, 49, 58, 69, 76, 79, 116, 131
functions, 9, 11, 12, 27, 55–57, 63, 66, 70, 77, 80, 83, 89, 92, 98, 101, 115, 130, 131, 139, 140, 143, 144

game, 21, 31, 34–38, 91
games, 4, 9, 22–24, 26, 35–38, 60, 129

gaming, 29, 34–37, 52
geometric, 11–14, 16, 20, 21, 27, 43, 46, 47, 49, 50, 56, 58–60, 62–66, 68, 69, 76, 78, 82, 84, 92, 94, 116, 117, 127, 128, 130, 133–135, 140, 143, 147
GLfloat, 63
GPS, 30, 36, 38
graph, 12, 13, 27, 87, 88, 91–97, 113
graphic, 3, 4, 12, 13, 21, 24, 27, 31, 33, 42–44, 49–51, 56–59, 62, 78, 86, 88, 116, 117, 132, 135
Graphics, 5, 9, 11, 22, 27, 38, 39, 86, 87, 101, 113, 128, 129
graphics, 3–6, 8, 11–13, 17, 18, 20, 22–30, 32–34, 36–38, 41–44, 46, 52, 55–58, 69, 82, 83, 86, 87, 117, 128–132
group, 19, 24, 30, 86, 91–93, 95, 100, 108, 109, 112, 138
guides, 29, 31, 33, 38, 52

handheld, 4, 7–9, 27, 33, 34, 40
handhelds, 3, 4, 6–8, 23, 26, 27, 38
hardware, 4, 5, 7, 10, 23–27, 32–34, 37, 39, 41–44, 46, 47, 49, 55, 56, 60, 82, 88, 130

Image2D, 91
input, 5, 8, 10, 12, 14, 27, 37, 38, 48, 56, 58, 59, 62–64, 66, 68, 70, 71, 74, 75, 83, 106, 117, 129, 130, 132–134, 139, 140, 142, 143, 145, 148
integer, 43–46, 59, 78, 106, 120, 133
interact, 3, 5, 11, 30, 90, 116, 124
interactive, 9, 16, 24, 34, 39, 86
interfaces, 4–12, 23, 29, 36, 57, 90, 115, 116, 127, 129, 131, 133

J2ME, 8, 25, 86, 90, 101
Java, 6, 7, 11, 22, 25–27, 38, 86, 87, 100, 113
JSR-184, 25–27, 86–88, 92, 95

keyboard, 8, 59, 108, 133, 134

languages, 7, 8, 11, 36, 44, 82
libraries, 11, 31, 34, 52, 56, 57, 113, 115, 128–130

library, 11, 22, 27, 38, 55–57, 64, 69, 84, 87, 113, 115–118, 123, 127, 128, 130, 131, 149
light, 5, 14, 20, 21, 33, 34, 68, 69, 71, 73, 74, 78, 81, 94–96, 104, 105, 117, 118, 125–127, 129, 142, 145, 147–149
lighting, 12, 14, 20, 40, 43, 49, 68–70, 72, 74, 75, 77, 95, 96, 117, 125–127, 142–144, 146, 147
lights, 20, 38, 68, 70, 72, 74, 87, 91–96, 103, 104, 125–127, 143, 147
line, 11, 19, 60, 61, 99, 100, 112
lines, 11, 16, 19, 20, 27, 47, 59, 60, 87, 88, 92, 116, 117, 119, 120, 144
link, 36, 57, 91, 130
linked, 91, 92, 97, 108, 115, 133
load, 43, 76, 94, 104
loading, 27, 76, 82, 88, 92, 116, 118
local, 94, 98, 121, 123

M3G, 38, 86–88, 91, 93, 100, 113, 115, 128
manipulation, 9
map, 29–31, 34, 39, 56
mapping, 21, 24, 27, 32, 76–79, 85
market, 4, 6, 8, 22, 23, 27, 35–38, 88, 128
material, 14, 15, 20, 22, 68–70, 73, 74, 96, 142–144, 147, 148
matrices, 59, 62, 83, 121
Matrix, 124
matrix, 59, 62, 63, 66, 68, 94, 109, 121, 123, 124, 140
matrixes, 47
memory, 4, 7, 14, 15, 20, 24, 26, 39, 41–43, 49–51, 55, 64, 76–78, 83, 86, 92, 118
mesh, 96, 97, 104, 107
Microsoft, 7, 115, 119, 127, 128
MIDlet, 89, 90, 101, 103
MIDP, 86, 88–91
Milgram, 32, 33
Mobile, 4–7, 11, 22, 26, 35, 38, 42, 44, 52, 86, 88, 101, 113, 115, 116, 127, 128
mobile, 3–9, 22–31, 33–43, 48, 52, 55, 60, 86, 87, 91, 113, 115–117, 128, 129
model, 8, 12, 13, 16, 18, 20, 23, 27, 29–31, 45, 59, 68, 69, 92, 95, 100, 104, 108, 116, 117, 121, 124, 126, 138, 139, 147
modeling, 12, 13, 16
module, 82
multimedia, 3, 6, 8, 39, 40, 52

navigation, 30, 31, 33, 37, 39
node, 87, 91–97, 99
nodes, 91–95, 97, 108
Nokia, 7, 24, 36, 38

object, 9, 12, 13, 18, 20, 33, 46, 56, 68, 69, 79, 81, 87, 91–100, 103, 104, 106, 107, 109, 111, 112, 115, 116, 120, 122–124, 128, 140
objects, 9, 12, 16, 18–20, 23, 27, 29–33, 39, 43, 46, 47, 58, 59, 78, 79, 81, 83, 91–95, 97, 100, 103, 106, 112, 115, 116, 121, 123, 124, 126, 127, 132, 140
OpenGL, 5, 13, 14, 22–27, 55–58, 68, 79, 82, 83, 87, 95, 121, 129–132
operating, 7, 27
operations, 14, 15, 27, 39, 41, 43, 46, 49, 55–57, 59, 60, 64, 69, 75, 78, 81–85, 95, 99, 117, 123, 124, 127, 130, 131
options, 116, 117, 120, 132
origin, 10, 68, 98, 99, 121, 123, 124, 137, 138, 140
orthographic, 59, 60, 63, 64, 136
OS, 7, 8, 25, 38, 57, 115, 130, 131
output, 8, 12, 15, 27, 40, 47, 56, 116, 117, 130

packages, 7, 8, 11, 12, 27
Palm, 7
parameter, 14, 49, 58, 59, 63, 64, 68, 70, 77, 78, 80, 83, 94, 95, 100, 108, 109, 112, 118, 121, 132, 133, 140, 142–144
parameters, 16, 58, 59, 62–64, 66, 68–71, 77–79, 83, 97, 106, 116–118, 120, 127, 129, 132–134, 139, 140, 142, 143, 145
PDA, 29, 30, 33, 39, 40
PDAs, 5, 34, 35, 37, 39, 52, 55, 86

pen, 4, 5
per, 24, 35, 41, 48, 63, 76, 77, 97, 125
per-fragment, 14, 15, 69, 78, 79, 85
per-pixel, 14, 15, 49, 82, 84, 97
per-vertex, 14, 15, 43, 60, 64, 84, 127
performance, 24, 27, 34, 37, 38, 41, 43, 47, 58, 128, 129
perspective, 12, 17, 18, 31, 59, 60, 76, 95, 97, 104, 122, 129
phones, 3, 5, 6, 8, 25–27, 29, 35, 43, 52, 55, 86
physical, 9, 87, 117
pipeline, 13, 14, 25–27, 46, 49, 50, 56, 59, 60, 78, 82, 84, 85, 92, 116, 117, 126–128
pixel, 14, 15, 20, 21, 24, 41, 43, 46–48, 50, 60, 75, 77, 117, 118
pixels, 14, 15, 19, 21, 41, 46–48, 50, 51, 75, 76, 79, 98, 116–118, 124
platform, 8, 22, 27, 84, 86, 118, 129
platforms, 5, 7, 8, 25, 26, 34, 38, 129
players, 32, 34, 36
playing, 36–38, 99
point, 11, 12, 18, 20, 24, 25, 30, 31, 33, 42–46, 48, 55, 59–61, 66, 77, 81, 83, 91, 94, 95, 112, 120, 121, 124, 126–129, 134, 139, 149
pointer, 58, 59, 64, 83, 107, 112, 116, 118, 133
polygon, 16, 21, 46, 68–70, 143
polygons, 11, 15, 46, 69, 76, 96
position, 9, 11, 12, 30, 40, 44, 59, 61, 73, 74, 95, 97, 99, 107–109, 118, 120, 124, 127, 147, 148
power, 3–5, 9, 23, 24, 27, 32, 36, 37, 39, 41, 42, 44, 49, 55, 78, 86, 129
primitive, 14, 20, 50, 51, 59–61, 64–66, 68, 71, 76, 117, 118, 135, 137, 139
primitives, 11, 14, 27, 43, 46, 49, 56, 58–60, 62–64, 76, 84, 115–118, 121, 127, 130, 133, 134, 137, 140, 143, 145
processor, 20, 27, 41, 42, 49, 50, 82, 86
program, 46, 58, 82, 88, 100, 133, 134
programming, 7, 11, 22, 27, 43, 44, 46, 52, 115, 118, 119, 128, 129
programs, 4, 26, 27, 39, 56, 57, 59, 77, 88, 129–131, 133
property, 74, 98, 99, 102, 116, 117, 148
prototype, 29, 30

rasterization, 15, 41, 46, 49, 76
reality, 4, 9, 29, 31–33, 38, 52
removal, 17, 18
rendering, 8, 14–16, 20, 23, 24, 27, 30, 37, 38, 41, 43, 46, 48–52, 56, 59, 60, 69, 75, 76, 82–86, 92, 116, 117, 120–122, 124, 126–129, 137, 150
representation, 12, 16, 29, 31, 44–46
resolution, 4, 9, 31, 37, 41, 43, 46, 47, 76, 129
RGB, 68, 76, 81, 83, 96

scanning, 19, 20
scene, 11–21, 24, 26, 27, 31–33, 37, 38, 52, 63, 69, 72, 75, 77–79, 87, 88, 91–97, 103, 104, 106–108, 113, 117, 118, 120–125, 147, 150
screen, 5, 8, 10–13, 15, 21, 23, 30, 37, 39, 41, 46–49, 56–58, 60, 64, 76, 78, 91, 94, 98, 100, 103, 107, 108, 117, 120, 124, 131–134, 136, 139, 140, 142
services, 29, 39
shape, 11, 12, 16, 20, 43, 47, 60, 64, 66, 68, 134, 136, 139, 140
shapes, 11, 12, 15, 27, 59–61, 66, 78, 97, 121, 126, 134, 140, 142
skeleton, 97
sketch, 62, 64, 65, 69, 70, 72, 73, 75, 77, 83, 85
software, 5–8, 11, 23–25, 27, 34, 39–41, 43, 55, 75, 82, 86, 88, 89, 115, 116, 125, 149
source, 12, 20, 55, 56, 68, 69, 74, 79, 80, 95, 101, 102, 115, 121, 127, 142, 149
space, 4, 5, 9, 10, 12, 27, 29, 38, 59, 60, 62, 64, 65, 76, 92, 94, 97, 109, 116–118, 121, 123, 124, 126, 134
specification, 24, 25, 27, 55, 84, 86, 88, 90, 95
specular, 20, 72–75, 117, 118, 125, 126, 147, 150
square, 41, 47, 60, 61, 63, 64
standard, 6, 8, 23–25, 27, 34, 36, 38, 55–57, 82, 83, 87, 88, 90, 91, 95, 107, 112, 113, 116, 121, 130

standards, 6, 22, 25, 26, 36
storage, 4, 5, 7, 39, 41
structure, 8, 12, 13, 34, 56, 78, 87, 92, 93, 96–98, 113, 123–125
study, 8, 30, 33, 47–49
support, 7, 9, 10, 24, 25, 29, 36–38, 43, 44, 83, 86, 87, 115, 129
surface, 16, 18, 20, 21, 38, 68, 69, 71, 78, 96, 122, 127, 145
surfaces, 12, 16–20, 46, 49, 56, 71, 81, 145
Symbian, 7, 8, 24, 25
synthetic, 32, 129
system, 6, 7, 9, 12, 13, 16, 25, 27, 31–33, 39, 46, 48, 57, 66, 68, 79, 83, 94, 98, 99, 116, 121, 123, 124, 126, 129–131, 140, 141
systems, 4, 7, 8, 27, 32, 33, 55, 127, 130

table, 36, 48, 49, 60, 118
tasks, 9, 11, 37, 40, 43, 44
technologies, 25, 33–35, 37, 86, 128
technology, 3, 5, 29, 31, 32, 35, 38, 41, 52, 115, 116
Texture, 21, 76, 82, 97
texture, 21, 24, 27, 32, 41, 46, 48, 76–79, 85, 91, 94, 96, 117, 129
textures, 15, 16, 41, 43, 46, 49, 76–78, 83, 106, 108
tile, 49–52
tiles, 49–51
torch, 74, 75, 148, 149
touch, 5
transformation, 11, 13, 43, 49, 59, 65, 66, 68, 76, 94, 97, 108, 109, 121, 123, 124, 137, 138, 140
transformations, 11, 14, 18, 27, 43, 56, 59, 62, 64, 66, 68, 82, 92, 94, 95, 104, 108, 109, 121, 124, 127, 128, 130, 140, 143
tree, 12, 13, 87, 93
triangle, 43, 49, 59, 61, 62, 64–66, 68, 69, 79, 120, 121, 136, 137
triangles, 15, 20, 43, 46, 49, 59, 61, 64, 65, 116, 117, 121, 134, 137

unit, 7, 42, 43, 98
uploaded, 76–78
Usability, 8, 30, 37, 39, 52
usability, 4, 8, 29, 37, 52
user, 4–9, 11, 23, 29–33, 37, 38, 59–61, 89
users, 3, 5, 8–12, 25, 29–31, 36, 85, 103

VBO, 83
VBOs, 83
vector, 24, 69, 71, 93, 98, 111, 118, 127, 145
vertex, 19, 20, 43, 59–64, 68, 69, 71, 78, 82, 83, 85, 97, 116–122, 125–127, 145
vertexes, 14–16, 19, 43
Vertices, 60, 116, 118, 134
vertices, 59–61, 63, 64, 66, 75, 76, 79, 83, 97, 116–122, 125–127, 134, 137, 139
video, 3, 5, 6, 9, 15, 22, 32, 34, 35, 38, 39, 57, 60, 76–78, 131
virtual, 4, 9, 12, 30–34, 36, 90
visible, 12, 16, 19, 49, 50, 78, 81, 94
visual, 4, 5, 9, 10, 39, 41, 46, 48, 57, 63, 72, 83, 147
visualization, 3, 4, 9, 29, 31, 34, 91, 104
visualizing, 76, 92, 104, 108
VR, 32
Vvrtices, 118

WAP, 37
Web, 37
window, 7, 27, 56–59, 62, 63, 68, 124, 130–135, 140, 142
wire-frame, 15, 16, 21, 32
wireframe, 16
wireless, 5–7, 9, 34, 38, 100, 128, 129
World, 32, 33, 92, 93, 107
world, 12, 20, 31–33, 36, 87, 92, 121, 123, 124, 128

RECEIVED
JUN 1 2007
ENGINEERING LIBRARY